Optimization of Manufacturing Systems Using the Internet of Things

Optimization of Manufacturing Systems Using the Internet of Things

Yingfeng Zhang
Fei Tao

ELSEVIER

AMSTERDAM • BOSTON • HEIDELBERG • LONDON
NEW YORK • OXFORD • PARIS • SAN DIEGO
SAN FRANCISCO • SINGAPORE • SYDNEY • TOKYO
Academic Press is an imprint of Elsevier

Academic Press is an imprint of Elsevier
125 London Wall, London EC2Y 5AS, United Kingdom
525 B Street, Suite 1800, San Diego, CA 92101-4495, United States
50 Hampshire Street, 5th Floor, Cambridge, MA 02139, United States
The Boulevard, Langford Lane, Kidlington, Oxford OX5 1GB, United Kingdom

Library of Congress Cataloging-in-Publication Data

A catalog record for this book is available from the Library of Congress

British Library Cataloguing-in-Publication Data

A catalogue record for this book is available from the British Library

ISBN: 978-0-12-809910-0

For information on all Academic Press publications
visit our website at https://www.elsevier.com/

Publisher: Joe Hayton
Acquisition Editor: Brian Guerin
Editorial Project Manager: Edward Payne
Production Project Manager: Caroline Johnson
Designer: Victoria Pearson

Typeset by Thomson Digital

Contents

Preface

The manufacturing industry produces wealth, and it is one of the significant elements of the sustainability of the modern society. The complexity of products is increasing while their life cycles become shorter. Thus, the traditional manufacturing systems and their related management approaches are required to be improved to adapt to the aforementioned changes. Internet of things (IoT) provides the potentiality to collect and communicate real-time data within the manufacturing systems, thereby achieving dynamic optimization and control of production. It helps to develop a more intelligent manufacturing system with higher flexibility and transparency.

In this book, the authors apply the IoT technology to manufacturing system (IoT-MS) to capture manufacturing data actively. Based on sufficient data, real-time system monitoring and the optimization of the modern dynamic manufacturing systems could be achieved. It is expected that this research work could contribute to the cutting-edge development of modern intelligent manufacturing systems.

This work is a summary of the authors' research works on the applications of IoT technology in manufacturing since 2010. The book includes10 chapters. Chapter 1 describes the newly advanced manufacturing technologies and intelligent manufacturing system, and then presents the conception of IoT and IoT-MS, and the challenges of IoT-MS. Chapter 2 proposes an overview of IoT-MS including the architecture, worklogic, and relevant core technologies. Chapter 3 describes the model and method of real-time and multisource manufacturing information perception. Chapter 4 presents the framework and the corresponding method of IoT-enabled smart assembly station. Chapter 5 describes the method and algorithm of cloud computing based manufacturing resources configuration. Chapter 6 describes the new strategy and method for IoT-enabled smart material handling. Chapter 7 presents the models and methods for real-time key production performances monitor. Chapter 8 presents the new strategy and method for real-time information-driven production scheduling. Chapter 9 illustrates the IoT-MS prototype system through a demo. Chapter 10 summarizes the conclusions and points out the future trends.

The contents of this book were planned and organized by Prof. Yingfeng Zhang and Prof. Fei Tao. Mr. Wenbo Wang, Mr. Geng Zhang, Mr. Sichao Liu, Mr. Dong Xi, Mr. Chen Qian, and Mr. Shan Ren assisted in the preparation of useful materials for Chapters 3–9.

This work was supported in part by the National Nature Science Foundation of China (NSFC) Under Grants No. 51175435, No. 51522501 and No. 51475032, the Doctoral Fund of Ministry of Education of China under Grant 20136102110022, and the 111 Project Grant of NPU under Grant B13044, and Beijing Nova Program and Beijing Natural Science Foundation (No.4152032).

The authors would like to acknowledge the graduate students Mr. Teng Yang, Mr. Jianxue Xu, Mr. Wei Du, Ms. Miao Li, and others for their contributions to the aforementioned research projects. In addition, the authors would like to acknowledge the contributors of the references used in this book.

The authors would like to express their special thanks to China Science Press and journal publishers. Some of the material has been published in *IEEE Transactions on Automation Science and Engineering* (TASE), *IEEE Transactions on Industrial Informatics* (TII), *International Journal of Production Economics* (IJPE), *Computers & Industrial Engineering* (CIE), *International Journal of Production Research* (IJPR), *Journal of Intelligent Manufacturing* (JIM), *International Journal of Advanced Manufacturing* (IJAMT), and *International Journal of Computer Integrated Manufacturing* (IJCIM).

The methodologies, technologies, and applications of IoT in manufacturing area are experiencing explosive development. This enables the optimization of dynamic production process that has drawn increasing attention from researchers worldwide. Hence, one of the purposes of the authors to deliver this book is to provide a platform to communicate with other researchers. The authors welcome any comments and suggestions.

Yingfeng Zhang
Fei Tao
Xi'an, July, 2016

Chapter 1

Introduction

1.1 THE CONCEPT OF IoT

As one of the most important new information technologies, Internet of things (IoT) has attracted great attention from governments, industries, and academia. For instance, the US National Intelligence Council has considered it as one of the six "Disruptive Civil Technologies" with potential impacts on US national power [1]. In 2009, European Union issued a report named "Internet of Things—An action plan for Europe." This report proposed that it is essential to take measures to ensure that Europe plays a leading role in the construction of new Internet [2], namely IoT. Besides, IoT was also officially listed as one of the five emerging strategic industries by Chinese government in 2010.

Originated from the radio frequency identification (RFID) system, it was first proposed by professor Ashton of MIT Auto-ID Labs in 1999. In 2001, Christopher in University of California proposed the concept of Smart Dust [3]. In 2003, Metro opened the first "future store" [4]. In 2005, Wal-Mart announced that the 100 largest retail stores would start to use RFID tag uniformly. In the Tunis World Summit on Information Society in 2005, the International Telecommunications Union (ITU) extended concept of IoT largely in the report of "ITU Internet Reports 2005 Executive Summary: The Internet of Things." As the explanation by ITU, it means the intelligent connectivity for anything at anytime and anywhere [5].

After 10 years of development, the new generation of information technologies has been highly integrated with IoT. In this context, the meaning of IoT is also constantly changing. Therefore, so far there is no clear and uniform definition about IoT, especially its various application backgrounds. It syntactically is composed of two terms which are "Internet" and "Things." Therefore, IoT can be understood from two perspectives, which are "Internet oriented" and "Things oriented" [6]. On one hand, from the perspective of "Things oriented," it refers to be based on standard communication protocols, and to form a worldwide network [7]. In other words, it is composed of a large number of things, which have identities and virtual personalities. In addition, these things are sustainable, enhanceable, and uniquely identified [8]. On the other hand, from the perspective of "Internet oriented," it can be considered as the expansion of Internet applications. As a light protocol, IP stack that already connects a huge amount of communicating device, have all the qualities to make IoT a reality [6].

Optimization of Manufacturing Systems Using the Internet of Things
http://dx.doi.org/10.1016/B978-0-12-809910-0.00001-9

In conclusion, based on "Internet," IoT extends and expands terminal of Internet to any objects and items. As a result, IoT constructs a network that covers everything in the world as well as the things in this network are able to exchange information and communicate with each other. Substantially, in order to achieve intelligent identification, positioning, tracking, monitoring, and management, everything in IoT is connected with Internet according to the arranged protocol by RFID, infrared sensor, global positioning system (GPS), laser scanner, and other information sensing equipment. Pervasive presence in IoT of a variety of things or objects, such as RFID tags, sensors, actuators, mobile phones, and so on are able to interact and cooperate with each other through unique addressing schemes to reach common goals [9] without human intervention. In addition, under the influence of related researches and developments, IoT has developed connection among different things to the combination and integration of information space and physical world.

The original architecture of IoT can be abstracted as perception layer, network layer, and application layer. However, the existing researches increasingly focus on ubiquitous service application of IoT. Therefore, according to the collection, transmission, processing, and application process of multisource information, the extended IoT architecture can be broadly summarized as perception layer, transmission layer, the cloud platform layer, and application layer, as shown in Fig. 1.1. The perception/sensing layer mainly achieves the ubiquitous object recognition and operation control. The transmission layer mainly realizes data acquisition and transmission. The cloud platform layer mainly conducts the corresponding information integration and data processing to support the ubiquitous service management and application. The application layer provides user interfaces for the applications of ubiquitous services of IoT combining related industries' demand.

At present, IoT has got good applications in a number of fields and industries. These applications can be grouped into the four domains [6]: (1) transportation and logistics domain, including logistics, assisted driving, mobile ticketing, monitoring environmental parameters, augmented maps, and so on, (2) healthcare domain, including tracking, identification and authentication, data collection, sensing, and others, (3) smart environment (home, office, and plant) domain, such as comfortable homes and offices, smart building, smart cities, industrial plants, smart museum, gym, and so on, and (4) personal and social domain, including social networking, historical queries, losses, thefts, and so forth.

1.2 EXISTING MANUFACTURING PARADIGMS AND THEIR LIMITATIONS

1.2.1 Agile Manufacturing

In 1991, a group of more than 150 industry executives participated in a study to meet the coming global challenges and revive the US manufacturing

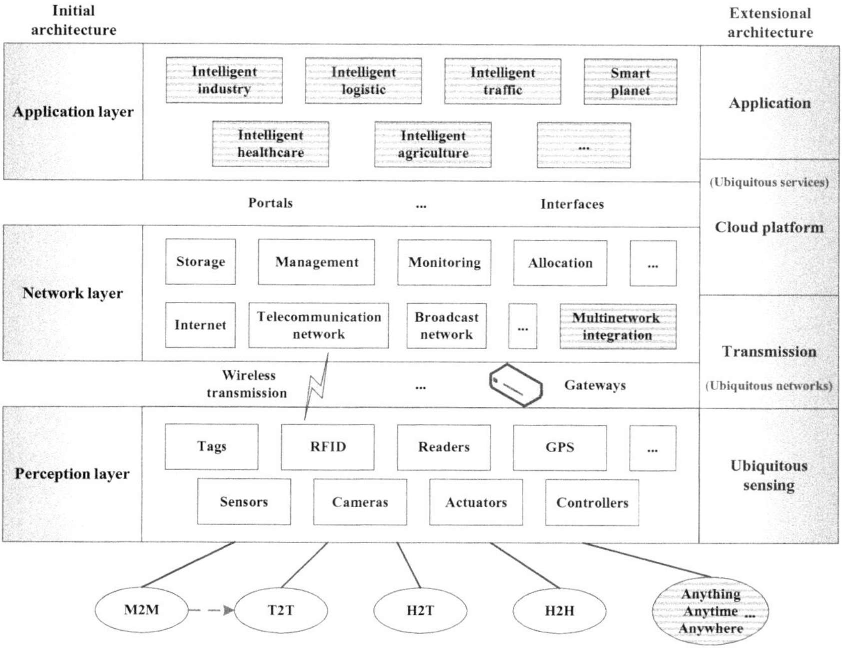

FIGURE 1.1 The general architecture of IoT.

treasures. Their efforts culminated in a two-volume report titled "21st Century Manufacturing Enterprise Strategy" [10]. As a result, the Agility Manufacturing Enterprise Forum (AMEF) affiliated with the Iacocca Institute at Lehigh University, was formed and the concept of agile manufacturing (AM) was introduced [11–14].

AM is an approach to manufacturing which is focused on meeting the needs of customers while maintaining high quality and controlling the overall costs involved in the production of a particular product. This approach is geared toward companies working in a highly competitive environment, where small variations in performance and product delivery can make a huge difference in the long term to a company's survival and reputation among consumers. The goal of AM is to keep a company stand out in the competition so that consumers will give priority to the company. Financial stability and strong customer support make it possible for the company to spend more on innovations. Companies which utilize an AM approach tend to have very strong networks with suppliers and related companies, along with numerous cooperative teams which work within the company to deliver products effectively. They can reconfigure facilities quickly, negotiate new agreements with suppliers and other partners in response to changing market forces, and take other steps to meet customer demands. This means that the company can increase the value of production with a

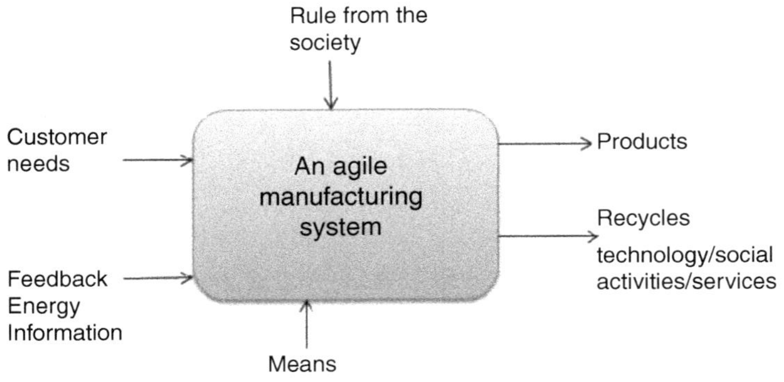

FIGURE 1.2 Principles of agile manufacturing (AM) [15].

high consumer demand, as well as redesign products to respond to issues which have emerged on the open market. The principles of AM are shown in Fig. 1.2.

The appearance of AM has prompted manufacturing enterprises to enter into the organizational mode of enterprise integration. In an environment with growing customized demands, computer network technology (CNT), supply chain management (SCM), and other related technologies have been developed rapidly. They enhance the agility, globalization of manufacturing operations. This period witnessed the rapid development of mass customization with the aim to achieve the economy of scope. However, analyzing the concept of AM, several issues need to be investigated, including how to acquire the need of customers, how to achieve on-demand use of various manufacturing resources, and how to establish the strong network with suppliers and related companies as mentioned previously. The technology of IoT and the strategies of on-demand use scheduling are the bottlenecks to achieve AM according to the issues.

1.2.2 Networked Manufacturing

Networked manufacturing (NM) means that manufacturing enterprises carry out product design, manufacturing, sales, procurement, management, and a series of activities based on network technology especially the Internet. It provides the environment and implementation methods for enterprises to develop remote collaborative design and manufacturing, online marketing, SCM, through the information integration, business process integration, and resource sharing among enterprises of different locations. NM can not only achieve the collaborations of product commerce, product design, product manufacturing, and supply chain, but also shorten the product development cycle, reduce research costs and improve the competitiveness of the whole industry chain [16–19], as shown in Fig. 1.3.

Technologies involved in NM can be divided into four parts, including overall technology, basic technology, integration technology, and application technology [21]. (1) Overall technology mainly refers to the technologies about the

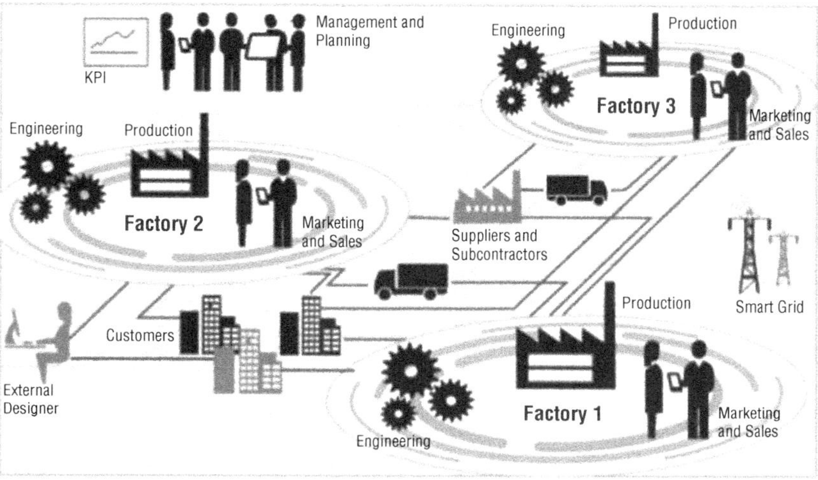

FIGURE 1.3 **Schematic of networked manufacturing (NM) [20].**

structure, management and application of NM system from the systematic point of view, for example, NM mode, the architecture of NM system, construction and organization methods, operation and management strategies, product life cycle management and collaborative product commerce technology. (2) Basic technology is the common and basic technology that is used not only in NM systems, but also in many other information systems, for example, basic theory and methods of NM, protocols and standardization technology of NM system, product modeling and enterprise modeling technology, workflow technology, multiagent system technology, virtual enterprises and dynamic alliance technology, and knowledge management and integration technology. (3) Integration technology contains the system integration and enabling technologies used in design, development, and application implementation of NM systems, for example, design/manufacturing resource and knowledge libraries technology, enterprise application integration technology, integration platform technology, ASP service platform technology, ecommerce and EDI technology, Web service technology, and so on. (4) Application technology refers to the technology involved in the application implementation of NM systems, for example, the approaches of NM application implementation, basic data library building and management technology, resource encapsulation and interface technology, cooperation technology in virtual enterprise, and network security technology.

However, in the practical application of NM, intelligent perception and connection of underlying physical manufacturing resources are the two key issues which hinder the development of NM. Although NM is developed on the basis of AM, the issue of how to achieve on-demand use is still unresolved. What's more, the low-degree servitization of manufacturing resource also is a bottleneck of NM.

1.2.3 Reconfigurable Manufacturing Systems

In 1995, researches of University of Michigan submitted a proposal to NSF for establishing a research center on reconfigurable manufacturing systems (RMSs) [22]. RMS is a new paradigm that attempts to link market demands and manufacturing systems by increasing flexibility of the system configurations. RMS is designed at the outset for rapid changes in structure, as well as in hardware and software components, in order to adjust production capacity and functionality quickly within a part family in response to sudden changes in market or in regulatory requirements [23].

As one of the components of RMSs, the reconfigurable machine tool (RMT) was invented in 1999 in the Engineering Research Center for Reconfigurable Manufacturing Systems (ERC/RMS) of University of Michigan [24]. The components of RMS are CNC machines, RMT, reconfigurable inspection machines and material transport systems (such as gantries and conveyors), which form the system. Different arrangements and configurations of these machines will have an impact on the system productivity. A collection of mathematical tools, which are defined as the RMS science base, may be utilized to maximize system productivity with the smallest possible number of machines.

Ideal RMSs possess six core characteristics, that is, modularity, integrability, customized flexibility, scalability, convertibility, and diagnosability [23–28]. These characteristics, which were introduced by Professor Yoram Koren in 1995, apply to the design of whole manufacturing systems, as well as to some of its components, including reconfigurable machines, controllers, and system control software, described as follows.

1.2.3.1 Modularity

In an RMS, all major components are modular (e.g., structural elements, axes, controls, software, and tooling). The benefits of modularity include economies of scale, increased feasibility of product/component change, increased product variety, reduced lead time, easier product diagnosis, maintenance, repair, and disposal [25,28].

1.2.3.2 Integrability

It is the ability to integrate modules rapidly and precisely by a set of mechanical, informational, and control interfaces enabling integration and communication. At the machine level, axes of motions and spindles can be integrated to form machines. Integration rules allow machine designers to relate clusters of part features and their corresponding machining operations to machine modules, which enables product–process integration. At the system level, the machines are the modules that are integrated via material transport systems (such as conveyors and gantries) to form a reconfigurable system. To aid in designing reconfigurable systems, system configuration rules are utilized. In addition, machine controllers can be designed for integration into a factory control system.

1.2.3.3 Customization

This characteristic drastically distinguishes RMS from flexible manufacturing systems (FMS), and has two aspects: customized flexibility and customized control. Customized flexibility means that machines are built around parts of the family that are being manufactured and provide only the flexibility needed for those specific parts, thereby reducing cost. Customized control is achieved by integrating control modules with the aid of open-architecture technology, providing the exact control functions needed [28].

1.2.3.4 Convertibility

It is the ability to easily transform the functionality of existing systems, machines, and controls to suit new production requirements [28]. Shorter conversion time between different production batches is a major requirement. To achieve this, the RMS must utilize not only conventional methods such as offline setting, but it should also contain advanced mechanisms that allow for easy conversion between parts, as well as sensing and control methods that enable quick calibration of the machines after conversion.

1.2.3.5 Scalability

It is the ability to easily change production capacity by rearranging an existing manufacturing system and/or changing the production capacity of reconfigurable stations [28]. Scalability is the counterpart characteristic of convertibility. Scalability may require at the machine level adding spindles to a machine to increase its productivity, and at the system level changing part routing or adding machines to expand the overall system capacity (i.e., maximum possible volume) as the market for the product grows.

1.2.3.6 Diagnosability

It is the ability to automatically read the current state of a system for detecting and diagnosing the root-cause of output product defects, and subsequently correct operational defects quickly [28]. Diagnosability has two aspects, including detecting machine failure and detecting unacceptable part quality. The second aspect is critical in RMS. As production systems are more reconfigurable, and their layouts are modified more frequently, it becomes essential to rapidly tune (or ramp-up) the newly reconfigured system so that it produces quality parts. Consequently, reconfigurable systems must also be designed with product quality measurement systems as an integral part.

In order to achieve RMS, several issues need to be investigated first, including how to link market demands and RMS, how to increase flexibility of the system configuration, and how to adjust quickly production capacity and functionality in response to sudden changes in market. However, related technologies to solve the aforementioned issues are relatively immature, for example, IoT technology, service-oriented technology (SOT), service management technology, on-demand use scheduling technology.

1.2.4 Product-Service System/Industrial Product-Service Systems

As industrialized countries are subject to a structural change toward a service society in which product services would meet users' demands, promote product innovation, be ecologically friendly, improve competitive strength, and be driven by new IT, product-service system (PSS)/industrial product-service system (IPS2) is developing rapidly as a new manufacturing system mode [29]. Services can be seen as add-ons to the actual product instead of the independent development in different departments. It links tangible products and intangible services, in order to change traditional manufacturing/consumption mode and solve environmental issues [30]. PSS is often defined as customers' life cycle–oriented combinations of products and services, realized in an extended value creation network, comprising a manufacturer as well as suppliers and service partners [31]. Similarly, IPS2 is sometimes characterized by the integrated and mutually determined planning, development, provision, and use of product and service shares including its immanent software components in business-to-business applications and represents a knowledge-intensive sociotechnical system [32]. The distinction between them is the integrated development of the mutually determined product and service shares is essential for IPS2. They all represent an integrated physical product and nonphysical service offering that delivers value in use. Essentially, PSS and IPS2 are usually mentioned as a similar conception.

The key research technologies of PSS/IPS2 could be classified into five categories as follows [33]. (1) Product service−oriented product design technology. Although there are great differences between different users for product service, some common demands among users exist. Therefore, modularization technology can satisfy the flexibility and reduce costs. At the same time, service-oriented reliability design, green design, and intelligence design should be taken into account. (2) Product service–oriented user demand mining technology. User demand could be mined through product service and social media. Furthermore, users can be involved in the enterprise innovation for the better understanding by enterprise. (3) Service-oriented product maintenance technology, such as remote diagnostics and maintenance, PMA, maintenance visualization, IETM, and knowledge service. (4) Product service–oriented information acquisition techniques, such as RFID and rapid measuring technology. (5) Product service–oriented user experience, such as an artificial emotion service robot and experiential marketing.

As the PSS/IPS2 develops, characters of product service differ from traditional ones. A holistic management approach as well as powerful methods and tools are compulsory for an IPS2 life cycle management. Existing product life cycle management requires several extensions for the better development, such as management of integrated product and services, dynamic customer-specific objects, interconnection of providers/customers, value-added processes, quality, and some other adaptions for different domains.

In a word, PSS/IPS2 leads the paradigm shift from traditional standardization and mass production into flexible and mass consumption which is driven by the additional value rather than price. However, there are some key issues that need to be investigated first, including how to acquire information about product and user efficiently and how to obtain users' demand accurately.

1.2.5 Manufacturing Grid

Manufacturing grid (MGrid) was first put forward in 2003 when information technology was experiencing the third wave of grid technology after the proliferation of the Internet and web technology [34]. Grid technology whose key concepts are resource sharing and coordinated problem solving in a virtual organization [35] can solve the very two bottlenecks that hinder the development of the current modes of manufacturing. In a grid system, users can obtain the services to fulfill their specific manufacturing requirements because of the connectivity and the operability among distributed and heterogeneous resources. Therefore, MGrid was put forward under this scenario while VM organizations based on grid technologies have gradually evolved to be a new manufacturing paradigm under a network-centric environment. In addition, some key technologies for implementing MGrid are adopted from computing grid.

Based on several related influential concepts of MGrid, the definition can be summarized as MGrid is to utilize grid, information, computer and advanced management, and advanced manufacturing technologies, etc. to overcome the barrier resulting from spatial distances in collaboration among different corporations to allow dispersed manufacturing resources (including design, manufacturing, human, and application system resources) to be fully connected, and it is a manufacturing service pool supporting manufacturing resource sharing, integration, and interoperability among different enterprises [36]. MGrid is indeed a realization of sharing various resources. Users can use all the resources distributed in heterogeneous systems and locations in MGrid as conveniently as they use local resources.

The key research contents and technologies of MGrid can be divided into the following four categories [37]. (1) The category of general technologies includes all the fundamental and necessary technologies such as concept, architecture, model, and management in manufacturing mode. (2) The category of supporting technologies includes the technologies supporting MGrid operation and system integration such as communication, security, and grid technologies. (3) The category of key enabling technologies include resource- and service-related technologies, management technologies, data and knowledge mining, payment, and evaluation. (4) The category of application technologies includes the interaction, visualization, cooperation, and portal-related technologies.

In addition, the related MGrid theories can be classified into the following categories. (1) Architecture of MGrid: different forms of MGrid architecture which decide the functions and basic modules have been proposed. (2) Resource management system: Resource management system is the central component of an MGrid system. (3) Modeling and encapsulation of MGrid resource: it is to enable MGrid resource information to be published and registered in an MGrid system. (4) Resource service discovery and scheduling: it emphasizes on search mechanisms or frameworks at the abstract level, resource match and search method from the descriptive information of resource service, service scheduling methods considering the factors of dynamic, intelligence degree and so on. (5) QoS management: it has been investigated from different perspectives of QoS whole-life cycle management, MGrid architecture view and QoS attribute parameter. (6) Service composition: MGrid system generates a new composite resource service and selects the optimal resource service composite path from all the possible paths to execute the task. (7) Workflow management: it has been widely investigated to enable users to compose and execute complex grid applications with distributed heterogeneous and unreliable computing resources. (8) Job management: it is primarily responsible for the whole-life cycle management of MGrid jobs that are anything that needs resource services. (9) Reliability management: reliability is a significant and complex issue in an MGrid system. (10) Security and trust management: the security problem is the key bottleneck that hinders the development and application of MGrid. (11) Others such as MGrid portal design, semantic support, and performance evaluation are also key theories of MGrid.

Although MGrid experiences a period of rapid development and research, some key issues such as embedded connect problems of physical manufacturing facilities, uniform protocols/standards and reliability problems still hinder the progress of MGrid.

1.2.6 Cloud Manufacturing

Cloud manufacturing (CMfg) is a computing- and service-oriented manufacturing model combining the emerged advanced technologies [e.g., cloud computing (CC), IoT, virtualization, and SOTs, advanced computing technologies (ACTs)] [38] with existing advanced manufacturing models (e.g., AM, NM, PSS/IPS2, MGrid) and enterprise information technologies. Users (e.g., provider, operator, and consumer) can easily have access to the manufacturing cloud services virtualized and encapsulated from the distributed manufacturing resources and manufacturing capabilities [39].

The first definition of CMfg was proposed in 2010, and then this new concept attracted much attention from both academic and industry communities. CMfg is a service-oriented, highly efficient and low consumptive, knowledge-based and intelligent networked AM model and technology, allowing manufacturing resources and capabilities to be virtualized and transformed into on-demand services available to users through a product life cycle [40]. Unfortunately, there still lacks a commonly accepted definition for CMfg, but some common terms, such as manufacturing resource, capability and platform, virtualization and cloud service are shared in different definitions. It is a promising manufacturing paradigm that is built on CC, IoT, cyber-physical systems (CPS), NM, service-oriented manufacturing, and virtual manufacturing.

Although existing research articles in this area present some system architectures that are have slightly different system structures, almost all of them share some similar elements, such as resource virtualization and cloud service composition. A popular hierarchical structure of CMfg system is divided into ten layers, as shown in Fig. 1.4. Resource layer, perception layer, resource virtualization layer, cloud service layer, application layer, portal layer, enterprise cooperation application layer, knowledge layer, cloud security layer and wider Internet layer (e.g., Internet, intranet, IoT). The details are shown in the Fig. 1.4. Core technologies (e.g., CC, IoT technologies, virtualization, SOTs, ACTs, existing manufacturing informationization technologies) will support the construction of the CMfg hierarchical structure.

Compared with existing manufacturing models, CMfg has several typical characteristics: service- and requirement-oriented, dynamic with uncertainty, knowledge-based initiative, physically distributed and logically centralized, Wikipedia style and group innovation-based manufacturing, lower threshold, and outsourcing. Based on the aforementioned typical characteristics, CMfg can offer several key advantages, such as (1) reducing resource idle capacity and increasing utilization, (2) reducing the up-front investments and lowering the cost

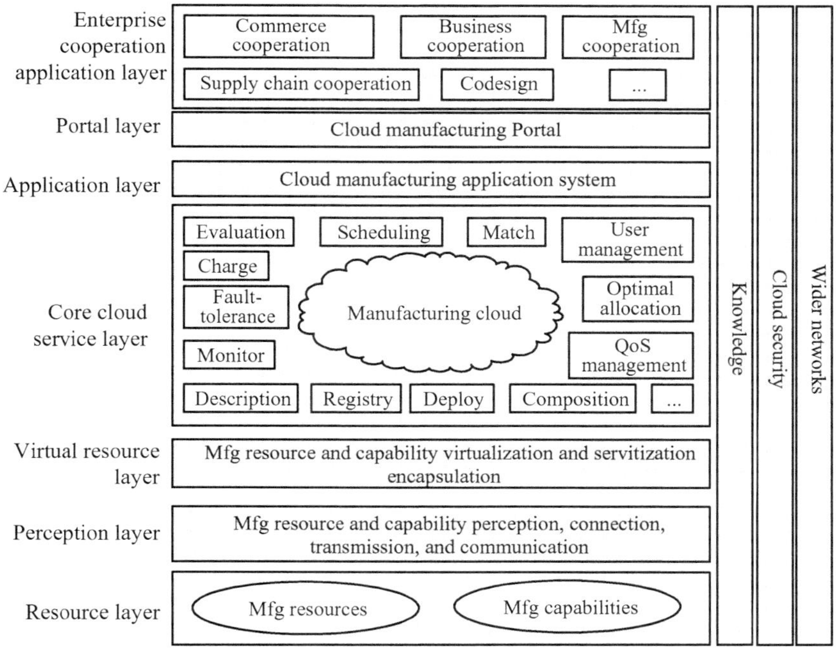

FIGURE 1.4 **Architecture of CMfg system [41].**

of entry for SMEs trying to benefit from high-value manufacturing resources (e.g., high-end equipment, expensive application systems) and specific manufacturing abilities that were hitherto available only to the larger corporations, (3) reducing infrastructure and administrative cost, energy saving, upgrades and maintenance cost, (4) easier for manufacturing enterprises (both smaller and larger) to scale their production and business according to client demand, (5) generating new types and classes of manufacturing/business model or process, (6) optimizing industrial distribution and speeding up the transformation from a distributed and high-energy consumption manufacturing model to a centralized, and environment friendly manufacturing model, and from production-oriented manufacturing to service-oriented manufacturing.

For implementing CMfg, intelligent perception and connection technologies for various physical manufacturing resources and abilities are the fundamental technologies. However, because of the diversity of physical manufacturing resources and capabilities, how to interconnect these different resources is still one of the most important challenges.

1.2.7 Limitations

After 20 years of development, these AMSs have played a very important role in developing modern manufacturing and industry, and in realizing goals of TQCSEFK (i.e., fastest Time-to-market, highest Quality, lowest Cost, best Service, cleanest Environment, greatest Flexibility, and high Knowledge). More and more AMSs devote to adapt to the trends and requirements of informatization, globalization, and servitization of manufacturing, and a lot of key technologies have been studied, including manufacturing resource and service modeling and encapsulation, resource and service optimal-allocation and scheduling, service workflow management, SCM, etc.

However, the socializations of the resources sharing, value creation, users' participation, supply–demand matching, on-demand use, and personalization in manufacturing run much clearer and faster as shown in Fig. 1.5. Due to the lack of common specifications and standards, the application of AMSs is hindered without realizing intelligent perception and connection of underlying physical manufacturing resources to Internet. As well, the low-degree servitization of MRs&Cs based on knowledge, limits the number, ability, and usage mode of services provided to users. In addition, due to the lack of effective operational mechanisms of resources and services and reliable safety solutions, it makes the application and development of the AMSs difficulty. Finally, the on-demand management of MRs&Cs is also the bottleneck to achieve the collaboration and intelligence of AMSs.

At the same time, SOTs, ACTs, virtualization technology, embedded technology, CPS, CC, IoT, have been quickly developed and widely applied. These new technologies have provided new methods for addressing the bottlenecks faced by the existing AMSs. For example, SOTs such as service-oriented architecture,

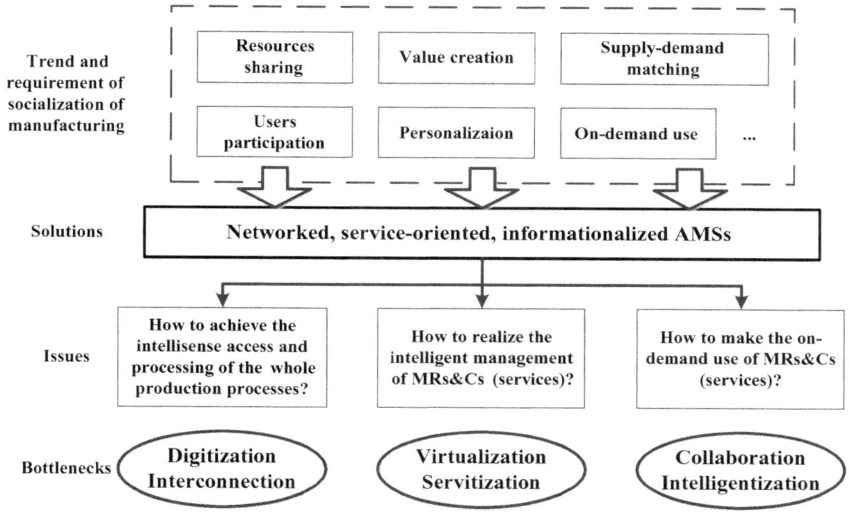

FIGURE 1.5 Bottlenecks of AMSs and the potential solutions [42].

web service, ontology, and semantic web, can provide enabling technologies for the construction of a virtual manufacturing and service environment. ACTs such as high-performance computing technologies, grid computing, parallel computing, can provide enabling technologies for solving complex and large-scale problems. Virtualization technology can hide the physical characteristics of the manufacturing resource and capability in an AMS from users. IoT technology can realize the effective connection, communication, and control from physical world to information world. Especially, as providing the new method for intelligent perception and connection of anything, and the on-demand use and efficient sharing of resources respectively, IoT and CC have been widely studied and applied in many fields. As a result, in order to realize the full sharing, free circulation, on-demand use, and optimal allocation of various MRs&Cs, it needs to investigate the applications of the theories and technologies of IoT in manufacturing at first.

1.3 APPLICATIONS OF IoT IN MANUFACTURING SYSTEM

So far, there still is no clear and uniform definition and architecture about IoT, it has been used in various application backgrounds. For instance, the applications of smart homes or smart buildings, smart cities, smart business, smart inventory and product management, healthcare, environmental monitoring, social security and surveillance, and so on [43]. Especially, it has been fundamentally changing the practical production and supply chain process and management with the aim of intelligent manufacturing.

Specifically, after introducing the generalized IoT into manufacturing industry, it can be devoted to address the "4Cs" (Connection, Communication, Computing, and Control) of MRs&Cs for the following different applications in manufacturing [42], as shown in Fig. 1.6. (1) Applications in the workshop. It can achieve the complete connection between terminal devices (i.e., various MRs&Cs) and the enterprise information management system for the automatic control of the IoT-enabled manufacturing execution in workshops. Therefore, three functions should be addressed: the access, identification, and control of the physical manufacturing execution process from materials and semifinished products to the final products. The data identified and acquired from the IoT-enabled manufacturing layer are the production- and product-related input of the enterprise information system. Moreover, the automatic control of manufacturing execution activities is the result under the output of the system to the PLC and other controllers. It is the general applications of IoT in workshops. (2) Applications in the enterprise. It promotes the integration of the production-related information, the product-related information, and other business management information, as well as the integration of the IoT-based workshop and other enterprise information subsystems. Enterprises can generate their own manufacturing services for the participation into the external supply chain, in addition to the management of the internal supply chain. It results in the origin of the local Internet of services (IoS). (3) Applications among enterprises. It addresses the information integration, storage, retrieval, analysis, use, data security, and other issues during these ubiquitous service management and application process among massive different enterprises. Consequently, CC and cloud platform technologies provide the new ideas and technical supports for the ubiquitous networking service management and application of IoT.

Besides, IoT has another important application in IoT-based energy management in product life cycle.

1. **Simulation and testing of product**
   Simulation and testing of product is a significant phase of product life cycle. Every small mistake during this process can lead to the abnormal product operation. One of the main challenges of this phase is to collect enormous amount of test data and analyze the data effectively. Nowadays, with the help of IoT, test data generated by IoT equipment is available. Then product designer and technical analysts can make energy-efficient decisions from the data connected with different kinds of influence factors.
2. **Public facilities real-time monitoring**
   Sensing terminals collect energy consumption information of water, electricity, gas, and heating. By comparing energy consumption for the same process in different environments and analyzing through the combination of quality of product, processing time, and energy-related data, the energy consumption condition of public facilities can be improved accordingly.

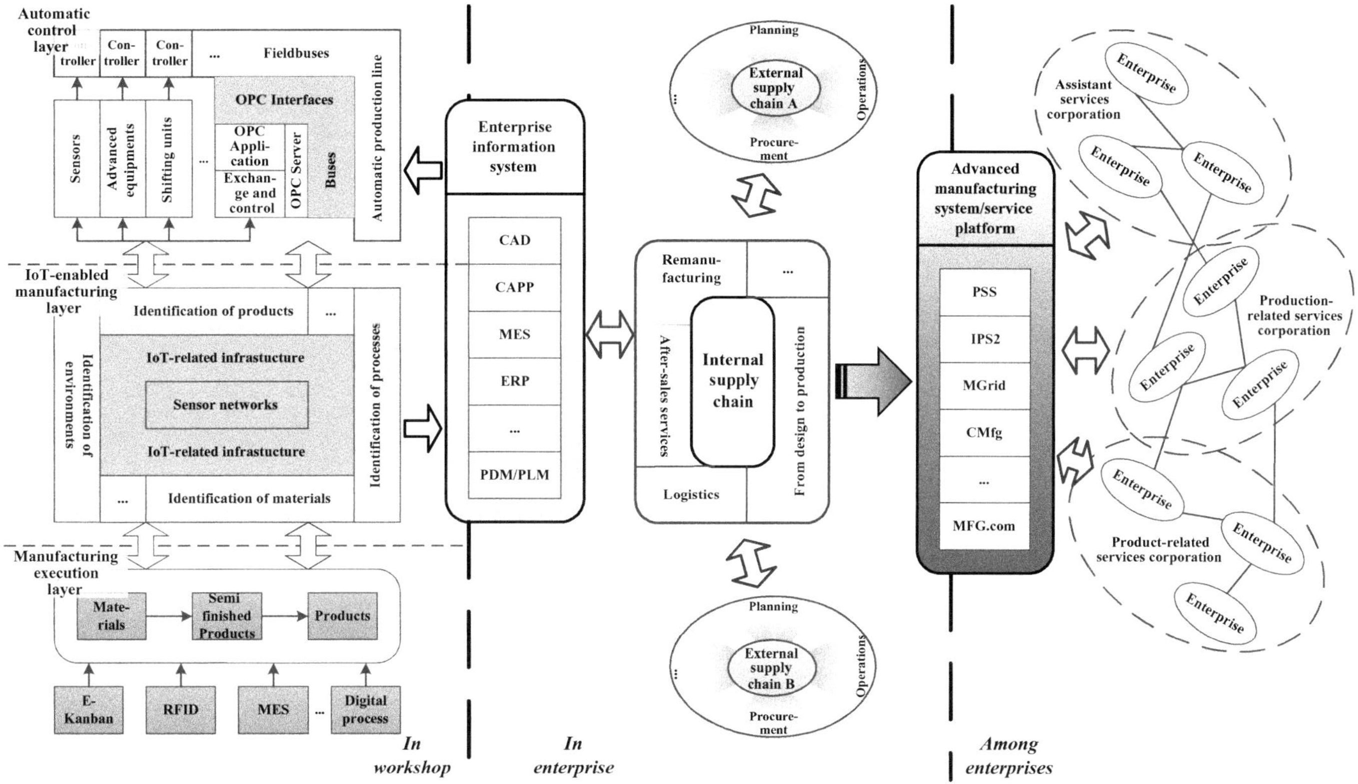

FIGURE 1.6 Applications of IoT in manufacturing [42].

3. Efficient working way of manufacturing equipment
The key to decrease the energy consumption in the manufacturing phase is improving the principles and configurations of equipment. With the help of IoT, equipment status and related parameters can be accurately collected. The power generation processes can be evaluated as well. Knowing equipment energy consumption pattern in real-time enables reducing energy consumption at peak time and reducing idle time by switching a machine off. Defining energy consumption for a machine in different configurations and choosing a more efficient machine configuration also decrease the energy consumption. In addition, as the manufacturing process is monitored in real time, the unqualified semifinished products can be discovered at once to avoid more energy waste.
4. Efficient transport planning
With the IoT, transport networks and product location tracking can be realized easily so as to improve energy efficiency in transportation phase. For example, "Energy Cards" have been used to determine and reduce energy demand in the logistics area. And some companies share transportation with other companies which share the same destination or transportation route to deliver products in order to achieve higher energy efficiency.
5. Energy conservation use guide
In utility phase, energy consumption can be easily generated by inappropriate behaviors of customers. Nowadays, with the help IoT technology, a large amount of data generated from the customer utility can be accurately analyzed and transferred to make correct decisions and give energy conservation advices for customers.
6. Automatic product inspection
In traditional mode, many products don't have regular inspection, which lead to potential energy consumption. With the support of IoT, product status is monitored continuously by IoT equipment. Thus, product inspection is taken automatically and continually during the utility phase. Undoubtedly, product inspection reduces general energy consumption.
7. Predictive and cooperative maintenance
Based on the continuous flow of information and global Internet platform, product maintenance can be enhanced to a predictive stage on a global basis. By comparing product's performance through globally networked monitoring systems, prediction of product abnormality, fault diagnostics, product degradation patterns analysis will enhance so that the product will be better prevented from unexpected breakdown in a ubiquitous way. In addition, when a fault occurs, proactive maintenance can be executed more efficiently and more cooperatively through the global platform.
8. Predicting remaining lifetime of parts or components
In most cases, while a product cannot be used, different level components still have its remaining value. Predicting remaining lifetime of parts or components helps to decide what to recycle and what the recycle level should be.

IoT technology helps predict the occurrence of breakdown and the relatively accurate remaining lifetime of parts or components by analyzing history data combining with corresponding component ID. This can largely reduce energy consumption in recycle phase.

In addition, Xu et al. [44] has reviewed the advances of IoT in industries, Bi et al. [45] studied the application of IoT in modern enterprise systems, Fan et al. [46] studied IoT-based smart rehabilitation system, and He et al. [47] have researched the application of IoT in the development of vehicular data cloud service.

At present, in manufacturing field, IoT technology is rapidly developing under the support of RFID, sensors, smart technology, and nanotechnology, and is expected to promote interconnection of anything. Then, IoT is helpful to construct a platform for sharing and interconnecting all kinds of manufacturing resources. Coupled with the rapid development of embedded systems and technologies, it provides enabling technologies for realizing the intelligent embedding of physical terminal equipment and the interconnection of M2M (including man-to-man, man-to-machine, and machine-to-machine) in manufacturing.

1.4 THE CONCEPTION OF IoT-MS

For better understanding, the definition of IoT-MS will be given at first. IoT-MS is defined as a multisource and real-time manufacturing data-driven manufacturing system, covering procedures of monitor, decision and management from the production orders assigned to the required work in progress or products finished.

IoT-MS consists of two main parts. The first is hardware part, which is responsible for configuring the sensors to capture the multisource and real-time data of the various manufacturing resources by deploying the IoT technologies to traditional manufacturing system. The second is the software part, which wraps a series of decision models, algorithms, and middleware, and is used to monitor, analyze, control, and make decisions for the whole manufacturing system.

1.5 KEY FEATURES AND LIMITATIONS OF IoT-MS

In contrast to current manufacturing systems, the following key features could be seen in the proposed IoT-MS.

Introduce an easy to use and easy to deploy architecture and solution for implementing smart manufacturing in the whole manufacturing systems using the IoTs.

Design the smart framework and models for improving the intelligence of the bottom-level manufacturing resources such as smart station and smart trolley because they are the key to intelligent manufacturing system.

Develop a new decision strategy and method for real-time information-based production scheduling and internal logistics optimization, which can be directly applied to manufacturing system, for example, shop floor, for improving the efficiency.

Present a critical event-based real-time key production performances analysis model, which can be used to actively identify the real-time production exceptions.

Although the proposed architecture, new smart models, strategies, and methods have significant contributions for improving the intelligence and real-time optimization of manufacturing systems, limitations still need to be discussed.

The manufacturing enterprises will invest much money to purchase the IoT hardware and software for capturing the real-time and multisource manufacturing data.

The new technologies of IoT will require the employees to learn specialized knowledge and skills. It is a new challenge for employees.

1.6 ORGANIZATION OF THE BOOK

This book systematically introduces the overall architecture, new models, methods, and core technologies related to optimization of manufacturing systems by using the IoTs. It includes 10 chapters, which are organized as follows.

Chapter 1 describes the newly advanced manufacturing technologies and intelligent manufacturing system, and then presents the conception of IoT and IoT-MS, and the challenges of IoT-MS. Chapter 2 proposes an overview of IoT-MS including the architecture, work logic, and relevant core technologies. Chapter 3 describes the model and method of real-time and multisource manufacturing information perception. Chapter 4 presents the framework and the corresponding method of IoT-enabled smart assembly station. Chapter 5 describes the method and algorithm of CC-based manufacturing resources configuration. Chapter 6 describes the new strategy and method for IoT-enabled smart material handling. Chapter 7 presents the models and methods for real-time key production performances monitor. Chapter 8 presents the new strategy and method for real-time information-driven production scheduling. Chapter 9 illustrates the IoT-MS prototype system through a demo. Chapter 10 summarizes the conclusions and points out the future trends.

REFERENCES

[1] National Intelligence Council, Disruptive Civil Technologies—Six Technologies With Potential Impacts on US Interests Out to 2025—Conference Report CR 2008–07, April 2008. Available from: http://www.dni.gov/nic/NIC_home.html

[2] http://eur-lex.europa.eu/LexUriServ/LexUriServ.do?uri=COM:2009:0278:FIN:EN:PDF

[3] B. Warneke, M. Last, B. Liebowitz, et al. Smart dust: communicating with a cubic-millimeter computer, Computer 34 (1) (2001) 44–51.

[4] K. Kalyanam, R. Lal, G. Wolfram, Future Store Technologies and Their Impact on Grocery Retailing—Retailing in the 21st Century, Springer, Berlin, Heidelberg, 2010, 141–158.
[5] ITU Internet Reports, The Internet of Things, November 2005.
[6] L. Atzori, A. Iera, G. Morabito, The Internet of Things: a survey, Comput. Netw. 54 (15) (2010) 2787–2805.
[7] INFSO D.4 Networked Enterprise & RFID INFSO G.2 Micro & Nanosystems, in: Co-operation with the Working Group RFID of the ETP EPOSS, Internet of Things in 2020, Roadmap for the Future, Version 1.1, May 27, 2008.
[8] B. Sterling, Shaping Things—Mediawork Pamphlets, The MIT Press, Cambridge, 2005.
[9] D. Giusto, A. Iera, G. Morabito, L. Atzori (Eds.), The Internet of Things, Springer, New York, 2010.
[10] L.M. Sanchez, R. Nagi, A review of agile manufacturing systems, Int. J. Prod. Res. 39 (16) (2001) 3561–3600.
[11] J.H. Sheridan, Agile manufacturing: stepping beyond lean production, Ind. Week. 242 (8) (1993) 30–46.
[12] L. Struebing, New approach to agile manufacturing, Qual. Prog. 28 (12) (1995) 18.
[13] C.W. Richards, Agile manufacturing: beyond lean?, Prod. Inv. Manage. J. 37 (2) (1996) 60.
[14] R.N. Nagel, R. Dove, 21st Century Manufacturing Enterprise Strategy: An Industry-Led View, Diane Publishing, Bethlehem, 1991.
[15] http://www.civilserviceindia.com/subject/Management/notes/flexibility-and-agility-in-manufacturing-systems.html
[16] M. Mitsuishi, T. Nagao, Networked manufacturing with reality sensation for technology transfer, CIRP Ann. 48 (1) (1999) 409–412.
[17] F. Liu, J. Liu, Q. Lei, The connotation and research development trend of networked manufacturing, Proceedings of China Mechanical Engineering Annual Conference, Beijing, China, 2002, 22–27.
[18] J.Q. Yan, Digitised and network-based manufacturing, Ind. Eng. Manage. 1 (2000) 8–11.
[19] S.Z. Yang, B. Wu, C.H. Hu, T. Cheng, Networked manufacturing and enterprise integration, China Mech. Eng. 11 (2000) 45–48.
[20] http://www.slideshare.net/stephane_potier/open-standards-key-to-success-for-industry-40-40582587
[21] Y. Fan, C. Huang, Y. Wang, et al. Architecture and operational mechanisms of networked manufacturing integrated platform, Int. J. Prod. Res. 43 (12) (2005) 2615–2629.
[22] Y. Koren, A.G. Ulsoy, Reconfigurable manufacturing systems, ERC/RMS report #1, Engineering Research Center for Reconfigurable Machining Systems, Ann Arbor, 1997.
[23] Y. Koren, U. Heisel, F. Jovane, et al. Reconfigurable manufacturing systems, CIRP Ann. 48 (2) (1999) 527–540.
[24] M.G. Mehrabi, A.G. Ulsoy, Y. Koren, Reconfigurable manufacturing systems: key to future manufacturing, J. Intell. Manuf. 11 (4) (2000) 403–419.
[25] R.M. Setchi, N. Lagos, Reconfigurability and reconfigurable manufacturing systems: state-of-the-art review, Proceedings of the Second IEEE International Conference on Industrial Informatics, 2004, INDIN'04, 529–535.
[26] Z.M. Bi, S.Y.T. Lang, W. Shen, et al. Reconfigurable manufacturing systems: the state of the art, Int. J. Prod. Res. 46 (4) (2008) 967–992.
[27] M.R. Abdi, A.W. Labib, Grouping and selecting products: the design key of reconfigurable manufacturing systems (RMSs), Int. J. Prod. Res. 42 (3) (2004) 521–546.
[28] Y. Koren, M. Shpitalni, Design of reconfigurable manufacturing systems, J. Manuf. Syst. 29 (4) (2010) 130–141.

[29] P.Y. Jiang, Q.Q. Zhu, Product service systems and research perspective, Manuf. Automat. 30 (12) (2008) 10–17.
[30] Y. Shimomura, T. Hara, T. Arai, A unified representation scheme for effective PSS development, CIRP Ann. 58 (1) (2009) 379–382.
[31] E. Schweitzer, J.C. Aurich, Continuous improvement of industrial product-service systems, CIRP J. Manuf. Sci. Technol. 3 (2) (2010) 158–164.
[32] H. Meier, R. Roy, G. Seliger, Industrial product-service systems—IPS2, CIRP Ann. 59 (2) (2010) 607–627.
[33] X.J. Gu, X. Li, G.N. Qi, Y.J. Ji, R.Z. Tang, P.Y. Jiang, Theory and key technology of product service system, J. Zhejiang Univ. 43 (12) (2009) 2237–2243.
[34] Y.S. Fan, D.Z. Zhao, L.Q. Zhang, S.X. Huang, B. Liu, Manufacturing grid: needs, concept, and architecture, Proceedings of the International Workshop on Grid and Cooperative Computing, Shanghai, China, December 7–10, 2003, 653–656.
[35] I. Foster, C. Kesselman, S. Tuecke, The anatomy of the grid: enabling scalable virtual organizations, Int. J. High Perform. Comput. Appl. 15 (3) (2001) 200–222.
[36] F. Tao, Y.F. Hu, Z.D. Zhou, Study on manufacturing grid and its resource service optimal-selection system, Int. J. Adv. Manuf. Technol. 37 (9–10) (2008) 1022–1041.
[37] F. Tao, L. Zhang, A.Y.C. Nee, A review of the application of grid technology in manufacturing, Int. J. Prod. Res. 49 (13) (2011) 4119–4155.
[38] L. Atzori, A. Iera, G. Morabito, The Internet of Things: a survey, Comput. Netw. 54 (5) (2010) 2787–2805.
[39] L. Ren, L. Zhang, F. Tao, C. Zhao, X.D. Chai, X.P. Zhao, Cloud manufacturing: from concept to practice, Enterp. Inf. Syst. 9 (2) (2015) 186–209.
[40] B.H. Li, L. Zhang, S.L. Wang, F. Tao, W. Cao, X.D. Jiang, X. Song, X.D. Chai, Cloud manufacturing: a new service-oriented networked manufacturing model, Comput. Integr. Manuf. Syst. 16 (1) (2010) 1–7.
[41] F. Tao, L. Zhang, V.C. Venkatesh, Y.L. Luo, Y. Cheng, Cloud manufacturing: a computing and service-oriented manufacturing model, Proc. Inst. Mech. Eng. B 225 (B10) (2011) 1969–1976.
[42] F. Tao, Y. Cheng, L. Da Xu, et al. CCIoT-CMfg: cloud computing and Internet of things-based cloud manufacturing service system, IEEE Transac. Ind. Inform. 10 (2) (2014) 1435–1442.
[43] D. Miorandi, S. Sicari, F. De Pellegrini, I. Chlamtac, Internet of Things: vision, applications and research challenges, Ad Hoc Netw. 10 (7) (2012) 1497–1516.
[44] L. Xu, W. He, S. Li, Internet of Things in industries: a survey, IEEE Trans. Ind. Inform. 10 (4) (2014) 2233–2243.
[45] Z. Bi, L. Xu, C. Wang, Internet of Things for enterprise systems of modern manufacturing, IEEE Trans. Ind. Inform. 10 (2) (2014) 1537–1546.
[46] Y. Fan, Y. Yin, L. Xu, Y. Zeng, F. Wu, IoT based smart rehabilitation system, IEEE Trans. Ind. Inform. 10 (2) (2014) 1568–1577.
[47] W. He, G. Yan, L. Xu, Developing vehicular data cloud services in the IoT environment, IEEE Trans. Ind. Inform. 10 (2) (2014) 1587–1595.

Chapter 2

Overview of IoT-Enabled Manufacturing System

2.1 INTRODUCTION

The term Internet of things (IoT) was first proposed by Kevin Ashton [1]. It refers to uniquely identifiable objects (Things) and their virtual representations in an Internet-alike structure. In fact, the rapid developments and applications in wireless sensors, communication and information network technologies, such as wireless sensors, radio frequency identification (RFID), auto-ID, Bluetooth, Wi-Fi, GSM, and so on have created a new era of the IoT.

Currently, the growth of competitive market globalization and customer demand diversification have led to the increasing demand of agility, networking, service, green, and socialization of manufacturing systems. Moreover, network, IoT, information, and other advanced technologies and theories have been fast developed and widely applied in the field of manufacturing. In a word, with the increasing competition in the global marketplace, manufacturing enterprises have to strive to be responsive to business changes which have further impacts upon production goals and performance at the manufacturing execution system level. According to our investigation, many business problems manufacturing enterprises are facing now are caused by lack of timely, accurate, and consistent manufacturing data of bottom level manufacturing resources. The overall architecture, infrastructure, and solution need to be proposed and designed to achieve real-time monitor, analysis, optimization, and control of the manufacturing systems. Therefore, it is essential for manufacturing industry to upgrade its actively sensing and dynamic optimization capabilities with advanced manufacturing technologies, sensor technology, multidiscipline knowledge, etc. in terms of IoT hardware, auto-ID devices, intelligent algorithms and decision-making software, and so on.

Many advanced manufacturing systems and its architecture are proposed by many scholars. The typical ones are described as follows.

Flexible manufacturing (FM) [2] have the ability to process or produce different products and allow rapid changes between them. FM aim to handle the uncertainty in product demand knowledge, finite manufacturing capacity,

Optimization of Manufacturing Systems Using the Internet of Things
http://dx.doi.org/10.1016/B978-0-12-809910-0.00002-0

and random machine failures. Computer-integrated manufacturing (CIM) [3] is a manufacturing paradigm that uses computers to control the entire production process. This integration allows individual processes to exchange information with each other and initiate actions. The concept of agile manufacturing (AM) [4] was introduced by the Agile Manufacturing Enterprise Forum (AMEF). It is a new paradigm which responds to complexity brought about by constant change. Concurrent engineering [4] is a work methodology based on the parallelization of tasks, which is sometimes called simultaneous engineering or integrated product development (IPD). It refers to an approach used in product development. Green manufacturing [5] is a kind of sustainable manufacturing mode with the full consideration of resources consumption and environmental impact. Sustainable manufacturing [6] is defined as the creation of manufactured products that use processes that are nonpolluting, conserve energy and natural resources, and are economically sound and safe for employees, communities, and consumers. Manufacturing grid (MG) [7] is an integrated solution to support the share and integration of resources in enterprise and social and for the cooperating operation and management of the enterprises. Industrial product-service systems (IPS2) [8] are specified by comprehensively considered product and service shares. The IPS2 represents a new solution-oriented approach for delivering value in use to the customer during the whole life cycle of a product. In addition to these advanced manufacturing paradigms mentioned previously, other patterns were also widely studied by researchers, such as global manufacturing [6], dynamic alliance [3], networked manufacturing (NM), crowdsourcing, and so on.

After many years study and application, these advanced manufacturing systems have been playing a very important role in the development of modern manufacturing and industry. However, in real-life manufacturing systems, according to our investigation, the bottleneck problem of manufacturing optimization and production control lies in the real-time data capturing from shop-floor frontlines and the lack of the seamless dual-way connectivity and interoperability architecture among enterprise, shop floor, and bottom manufacturing resources such as machines, trolleys, and so on. As a result, the requisite information is either unavailable or behindhand. The laggard information transfer flow and the unmatched information transfer method enhance the difficulty for the up-level decision of the enterprises [9].

The research questions for developing and implementing real-time monitor, analysis, control, and dynamic optimization of the manufacturing systems could be summarized as follows:

1. The first research question is the overall information capturing, integration, and decision architecture to track, trace, and transmit the real-time manufacturing and monitor, analysis, control, and dynamic optimization among manufacturing system layer, shop-floor layer, and bottom manufacturing resources layer.

2. The second research question is the deploy IoT technologies such as RFID, auto-ID sensors to make the bottom manufacturing resources smart, so that the real-time status of the distributed manufacturing things such as operator, material items, pallets, and so on could be automatically captured.
3. The third research question is the sharing, exchanging, and integrating the real-time manufacturing information with heterogeneous enterprise information management systems.

Considering the advantages of the IoT, in this chapter, an overall architecture of the manufacturing systems using Internet of things (IoT-MS) is presented. It aims to provide a new paradigm for real-time monitor, analysis, control, and dynamic optimization of the manufacturing systems by extending the IoT technologies to manufacturing field. Under this IoT-MS architecture, the manufacturing things such as operators, machines, pallets, materials, and so on can be embedded with sensors; they can interact with each other. The changed information and their status could thus be tracked and integrated with heterogeneous enterprise management information systems. The proposed IoT-MS will facilitate the real-time information-driven active monitor, analysis, control, and dynamic optimization of the manufacturing systems.

The rest of the chapter is organized as follows. Section 2.2 reviews the literature of the architecture of advanced manufacturing technologies and systems, manufacturing information standard, and share and integration method. Section 2.3 presents the overall architecture of the IoT-MS. The real-time manufacturing information sharing and integration service is described in Section 2.4. Section 2.5 introduces the worklogic of IoT-MS. The core technologies of IoT-MS are given in Section 2.6.

2.2 RELATED WORK

In general, reviewing the published literature aims to evaluate the body of literature and identify potential research gaps highlighting the boundaries of knowledge [10]. In this section, we will not only provide a brief overview of past research on the topic of advanced manufacturing paradigms and technologies but also on the topic of the standard and method of manufacturing information share and integration. Our aim is to identify the most influential studies, determine the areas of current research interest and provide insights for current research interests and directions for future research in the IoT-MS.

To sum up, the research literature related to this chapter can be divided into two parts: (1) advanced manufacturing paradigms and technologies; and (2) manufacturing information standard and share and integration method.

2.2.1 Advanced Manufacturing Paradigms and Technologies

Over the past 20 years, manufacturing enterprises have been able to dynamically optimize their production processes and dramatically improve product quality and yield by implementing the advanced manufacturing paradigm (e.g., lean manufacturing, service-oriented manufacturing, and so on). However, in some manufacturing environments, for instance, metallurgy, chemicals, and mining, extreme swings in variability are a fact of life [11], sometimes even after the advanced manufacturing paradigm have been applied. Given the complexity of production activities that influence the decision making of manufacturing process in these and other industries, manufacturing enterprises need some advanced technologies to track, diagnose, and optimize the process flaws. Therefore, IoT, cloud computing (CC) and cyber-physical system (CPS) technologies now are widely used to develop the new manufacturing paradigms to provide aforementioned capability for manufacturing enterprises.

In the IoT paradigm, many of the entities and objects that surround us will be connected to the network. RFID and sensor network technologies have been raised to meet this new challenge, in which information and communication systems were invisibly embedded in the environment around us [12]. The earliest case of an industrial IoT application was the supply chain and logistics management. RFIDs can be attached to (or embedded in) objects and used to identify materials and goods [13]. By using the IoT technology during the process of product life cycle management (PLM), the entire lifecycle status of products can be tracked by RFIDs [14]. For example, RFID readers can be installed along the production plant to monitor the production process, while the label can be traced throughout the entire supply chain (e.g., packaging, transportation, warehousing, sale, service, maintenance, and disposal). Advanced IoT systems, consisted of RFID-equipped items and smart shelves, where the objects and products can be tracked in real time. This may help to reduce material waste, thus lowering costs and improving profit margins for both retailers and manufacturers [15]. IoT can be used to offer advanced solutions in the automotive industry [16]. Each parameter of the automobile can be monitored by specific sensors, such as tire pressure, motor data, fuel consumption, location, speed, and so on. The sensed data was then reported to the center system for optimization of product design, improvement of service, and energy efficiency. IoT may help to increase the environmental sustainability of our cities and the people's quality of life [17]. For example, smart parking systems may guide drivers to the nearest available parking slot according to the real-time data of drivers' location or real-time recommendation of the smart parking systems, hence saving time and fuel, and thus reducing the carbon footprint. Additional applications include smart services for entertainment and tourism [18]. For example, by taking pictures of monuments and other tourist landmarks, the users may obtain pertinent information on the personal smartphone from the recommendation system of the city tourism services center, and be guided to discover the heritage of the city. IoT can be used for monitoring and exchanging information about energy flows in the smart grid [19]. In the application of grid, smart meters, automatic control devices, smart switches, and smart appliances were used to monitor the real-time data of electric consumption. By intelligent data analysis, the grid was able to know in advance the expected demands and to adapt the production and consumption of electricity. Consequently, the peak loads of power consumption can be avoided, the possibility of a power outage can be eliminated, and the promptness of action in case of failure and fault can be enhanced. Other types of applications were integrated IoT with the smart grid to optimize the domestic consumption [20]. For example, the home area network (HAN) allows appliances to interact with smart meters in order to reduce costs while guaranteeing the demanded performance. Emergency management assists the society in preparing for, and coping with manmade or natural disasters such as chemical leaks, floods, fire, earthquakes, tornadoes, and epidemics [21]. IoT offers feasible solutions for monitoring and tackling these emergency scenarios in real time (or near real time). The medical and healthcare sector will be strongly affected by IoT [22]. For instance, body area networks (BANs) which is formed by smart and wearable devices (e.g., watches, shoes, and glasses), allow doctors to conduct the remote patient's monitoring and remote patient's diagnosing of the hospital.

A new service-oriented manufacturing paradigm, namely cloud manufacturing (CMfg) has been proposed to transform from production-oriented manufacturing to service-oriented manufacturing, and to construct complete manufacturing enterprise systems (ES) with high intelligence [23]. This new manufacturing paradigm was proposed with combination of advanced computing models and technologies such as CC [24], high performance computing, service-oriented technologies [25], and IoT. To fully implement CMfg, the concept, architecture, core enabling technologies, and characteristics of CMfg were widely studied [26–29]. A cloud-based design and manufacture (CBDM) model composed of a cloud consumer, cloud provider, cloud broker, and cloud carriers was studied [30,31]. Product configurators of CMfg for enterprises were studied by Yip et al. [32] to achieve product customization in order to address individual customers' requirements. Wang [33] studied and developed an Internet and web-based service-oriented system for machine availability monitoring and process planning toward CMfg. The problem of manufacturing resource and service management were studied by Tao et al. [23,34], including the utility modeling, equilibrium, and collaboration of manufacturing resource service transaction, and resource service scheduling based on utility evaluation, service composition and its optimal-selection algorithms, service composition network in CMfg system or service-oriented manufacturing systems. Lu et al. [35] used ontology within a cloud management engine to manage user-defined clouds. The ontology is used to instantiate companies on which the user executes a customized rule to create a cloud and define users authorized to access this cloud. With the emerging theory of "Industry 4.0," the integration of cloud technologies and industrial cyber-physical systems (ICPS) was first studied [36]. In this paper, the development and character of ICPS were described. With the support of the cloud, ICPS development will impact value creation, business models, downstream services, and work organization. A cloud-based production planning and control system for discrete manufacturing environments were developed to meet the shift of traditional mass producing industries toward mass customization practices [37]. The proposed approach takes into consideration capacity constraints, lot sizing, and priority control in a "bucket-less" manufacturing environment. Gao et al. [38] review the historical development of prognosis theories and techniques and project their future growth enabled by the emerging cloud infrastructure. Techniques for CC were highlighted, as well as the influence of these techniques on the paradigm of cloud-enabled prognosis for manufacturing. Virtualization was critical for resource sharing and dynamic allocation in CMfg. An effective method for manufacturing resources and capabilities virtualization was proposed by Liu and Li [39]. This method contains manufacturing resources modeling and manufacturing cloud services encapsulation. A manufacturing resource virtual description model was built, which includes both nonfunctional and functional features of manufacturing resources. The model can provide a comprehensive manufacturing resource view and information for various manufacturing applications.

Servitization plays an increasingly important role in modern manufacturing environment. In respect of service-oriented manufacturing paradigm, intangible services and physical products were integrated into one system, which was called product-service system (PSS) to provide a comprehensive solution for customers [40]. The ultimate PSS objective was to increase a company's competitiveness and profitability [41], and another of the PSS objectives was to reduce the consumption of products through alternative scenarios of product use instead of acquiring it. For example, customers who drive rather infrequently may not need to buy cars but would use a car-sharing system [42]. When defining the PSS, Goedkoop et al. [43] define it as a combination of products and services in a system that provides functionality for consumers and reduces environmental impact. Mont [44] highlights how the PSS offers a product and system of integrated products and services that are intended to reduce the environmental impact through alternative scenarios of product use. Thus, the PSS was a competitive opportunity, which was important for how it was able to alter consumption standards. In other words, this new manufacturing paradigm aims to improve both competitiveness and the pursuit of balance between social, economic, and environmental issues [45]. A RFID-enabled PSS for automotive part and accessory manufacturing alliances were proposed by Huang et al. [46] to alleviate the manufacturing systems of automotive part and accessory manufacturers, and to address the "three high problems," namely high cost, high risk, and high level of technical skills. To have a competitive PSS, a methodology of PSS engineering design was proposed to support engineering designers during the development process [47]. The context of PSS development and the current methods used to develop such systems were described in this paper. Then, the tools and formalism used in the proposed methodology based on a function-oriented description and an activities-related description were explained. An approach for PSS configuration was proposed to achieve desired benefits and customer satisfaction [48]. In this research, the customer needs were first divided as functional needs and perception needs which were generally expressed by customers in their own words. Based on it, a multiclass support vector machine model was built for configuring a specific PSS that meets customer needs. Since methods to evaluate the feasibility of new businesses vary with the characteristics of businesses, the evaluation methods might need to be modified to reflect unique nature in the design of a new PSS. To address this problem, a new framework was proposed to improve the applicability of evaluation methods of newly designed PSS [49]. Environmental constraints lead to important changes in the innovation strategies of manufacturing firms. The development of PSS in 10 manufacturing firms were studied by Laperche and Picard [50] to investigate the reasons and the forms of PSS development. Their impact on innovation management as well as the prerequisites and limits of their implementation were discussed in detail.

2.2.2 Manufacturing Information Standard and Share and Integration Method

In the real-world manufacturing environment, different enterprises usually use the different middleware and software applications. The problem of information integration should be considered and addressed when information was exchanged among heterogeneous enterprise information systems (EISs). Therefore, unified data models and standard data schemas of manufacturing information play key roles in information sharing and integration of heterogeneous EISs. This was not only at business or at manufacturing levels but also inside a single enterprise or between networked enterprises [51].

An interface to achieve seamless connections between enterprise resource planning (ERP) and process control systems was developed by Siemens Energy & Automation, Inc. [52]. Panetto and Molina [53] described the challenges, trends, and issues for the enterprise information interoperability. They pointed out that enterprise knowledge sharing, common best practices use, and open source/web-based applications were helping to achieve the concept of integrated enterprise and hence the implementation and interoperability of networked enterprises. Rodriguez et al. [54] presented a novel classification method for eight web service discoverability antipatterns, which was good for ranking more relevant services. Kong et al. [55] modeled the uncertain workload and the vague availability of virtualized server nodes with a fuzzy prediction method. The ISA95 standard [56] was developed with the objective to reduce the cost, risk, and errors associated with implementing interfaces between enterprise and production control systems. The business-to-manufacturing mark-up language (B2MML) [57] standard developed by World Batch Forum (WBF) specifies accepted definitions and data formats for information exchange between different management systems. Anaya et al. [58] described a unified enterprise modeling language (UEML), which aims at supporting integrated use of enterprise information models expressed using different languages. Zhang et al. [59] adopt extensible mark-up language (XML)-based schemas to implement the information integration of the heterogeneous information systems. A comprehensive machine tool resource model was proposed by Vichare et al. [60]. The proposed model considers the majority of standard machine elements including specification of the NC controller. Shi et al. [61] employed XML schema to encapsulate manufacturing resource information and adopted web service description language (WSDL) to model the accessing operations to manufacturing resources. Liu et al. [62] proposed a multigranularity resource virtualization and sharing strategies for bridging the gap between complex manufacturing tasks and underlying resources. The proposed approach considers three factors, including workflow, activity, and resource that significantly influence stepwise decompositions of a complex manufacturing task. Tao et al. [63] investigated the formulation of service composition optimal-selection in CMfg with multiple objectives and constraints. An extended product data model that can specify technical services, taking into account the product design and manufacturing process supported as integrated product life cycle data was proposed to maximize product performance over its life cycle [64]. Being an information modeling language to support the Standard for Exchange of Product Model Data (STEP), EXPRESS has been developed to share and exchange product design, manufacturing, and production data in product life cycle. In order to explicitly represent and handle fuzzy engineering information, an extended EXPRESS-G data model for different kinds of fuzziness modeling was developed [65]. The formal transformation from the fuzzy EXPRESS-G data model to the fuzzy XML model was investigated in this paper. The formal transformation approaches proposed in the paper were demonstrated with engineering application examples. The STEP standard makes it easier to integrate systems that process various product life cycle functions, such as design, engineering, manufacturing, logistics support, and will help to facilitate concurrent engineering. To facilitate the computer-readable exchange of the product bill of materials (BOM) information for product data management (PDM), the STEP ISO 10303-21 was implemented by Shih [66] in order to share the product data information in a manufacturing environment.

2.3 OVERALL ARCHITECTURE OF IoT-MS

Fig. 2.1 shows the overall architecture of IoT-MS. It aims to provide an easy-to-use and easy-to-develop solution for real-time monitor, control, and optimization of manufacturing system by using the IoT. As seen in Fig. 2.1, by applying the conception of IoT to manufacturing systems, the real-time status of distributed manufacturing resources such as manufacturing stations and trolleys can be easily sensed. Then, the real-time performance and exceptions of the whole manufacturing systems could be dynamically monitored, which will provide important support for implementing optimal control and decision making.

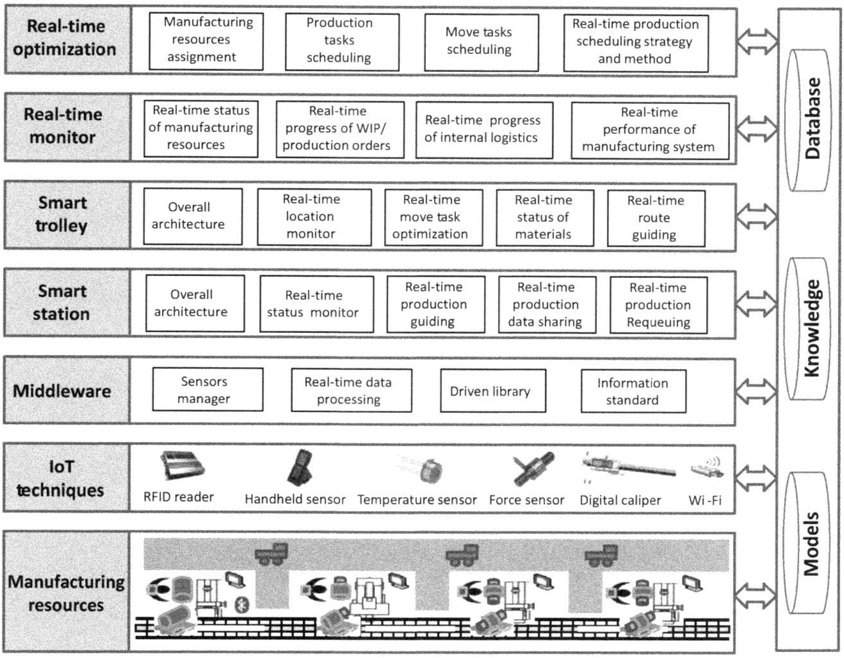

FIGURE 2.1 **Overall architecture of IoT-MS.**

From the bottom to top of Fig. 2.1, there are some key components designed for IoT-MS. They are briefly described as follows.

1. Manufacturing resources

   Generally, manufacturing resources are the typical resources in the manufacturing system, for example, manufacturing shop floor, machine, operator, trolley, enterprise management information systems, and so on.
2. IoT techniques
   IoT techniques are responsible for providing the hardware solution for sensing and capturing the real-time and multisource manufacturing information. Different types of sensors or auto-ID devices will be configured together with the manufacturing resources so that the real-time manufacturing data can be actively captured. For example, RFID readers are used to capture the real-time location data of operator, material, trolley, WIP item, and so on; digital caliper can be used to capture the quality data of the workpiece.
3. Middleware
   Because different types of sensors or auto-ID devices may have different driven and data structure, it is very important to design a middleware to centrally manage the heterogeneous sensors or auto-ID devices.

4. Smart station
   Manufacturing machines play an important role for executing the production tasks of the manufacturing system. Therefore, the smart model of machine sides are the fundamentals of the smart manufacturing system. In this book, the assembly station is selected as the bottom production object, and the smart station model is established. It aims to make the physic station smart with the capability of active sensing and self-decision.
5. Smart trolley
   Material handling play an important role for executing the internal logistics tasks of the manufacturing system. Therefore, the smart model of trolley sides are the fundamental of the smart manufacturing system. In this book, the trolley is selected as the bottom material handling object, and the smart trolley model is designed and established. It aims to make the physic trolley smart with the capability of intelligent navigation such as actively capturing a move task, navigating the trolley to load or unload the material items at right locations, and so on.
6. Real-time monitor
   The real-time traceability and visibility of manufacturing things plays a critical role in improving shop-floor performance with better planning, scheduling, and control decisions.
7. Real-time optimization
   The sensor networks are used to measure the dynamical parameters such as environment (temperature, pressure, humidity, etc.), movement (velocity, acceleration, shock, etc.) and real-time status of the manufacturing things. It will provide the supervisors with real-time information to implement optimal control and decision of a manufacturing execution system.

2.4 INTEGRATION FRAMEWORK OF REAL-TIME MANUFACTURING INFORMATION

2.4.1 Framework of Real-Time Manufacturing Information Sharing and Integration

As different manufacturing companies may adopt the different enterprise information management systems, in IoT-MS environment, there are two great challenges for real-time manufacturing information sharing and integration. The first is how to process a huge amount of real-time data captured from the distributed sensors to useful manufacturing information. The second is how to share and exchange the real-time manufacturing information among the heterogeneous enterprise information management systems such as ERP, MRP, PDM, MES, CAPP, and so on.

To deal with these problems, a real-time manufacturing information sharing and integration framework (RMISIF) is presented as seen in Fig. 2.2. It aims to build up a bridge to share and exchange real-time information between heterogeneous

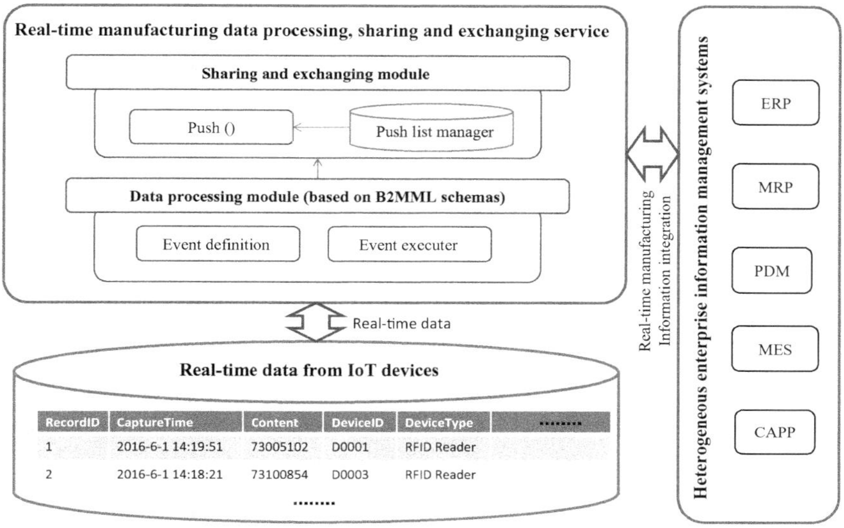

FIGURE 2.2 Framework of real-time manufacturing information sharing and integration.

enterprise information management systems and real-time manufacturing data captured by IoT devices. As shown in Fig. 2.2, the bottom shows the real-time manufacturing data database. The top shows the real-time manufacturing data processing, sharing, and exchanging service, and the B2MML standards are adopted to create standard schemas for sharing and exchanging among the heterogeneous enterprise information management systems (shown in the right of the Fig. 2.2).

2.4.2 Real-Time Manufacturing Data Processing, Sharing, and Exchanging Service

The real-time manufacturing data processing, sharing, and exchanging service (PSES) is the core of the manufacturing information sharing and integration framework. The inputs of PSES are the parameters of the data source of the heterogeneous enterprise information management systems which users want to acquire or update information from or to, while the outputs are to provide the standard real-time information for the heterogeneous enterprise information management systems.

PSES follows the service-oriented architecture and is represented as a web service which can be easily published, searched, and invoked through Internet. It includes two main modules, namely data processing module and sharing and exchanging module, which are described as follows.

1. Data processing module
   Data processing module is responsible for processing the isolated and inconsistent data sensed by IoT devices to standard information schema so that

it can be easily shared and exchanged among the heterogeneous enterprise information management systems.

To make the processed manufacturing information can be understood by many heterogeneous enterprise information management systems, this module uses the ISA 95 and B2MML data structure and schemas. Fig. 2.3 shows the overall of real-time manufacturing information schema (RMIS) based on B2MML. The designed RMIS includes and extends the set of standards and data structure of ISA 95 and B2MML.

As seen in Fig. 2.3, the model of RMIS is in accord with the hierarchy of a manufacturing system, namely a manufacturing system has one or more shop floors. Each shop floor consists of one or more production lines/stations. Each production line or station involves a variety of manufacturing things such as man, equipment, materials, WIP, and so on. Take the information schema of real-time process segment as an example, it has eight types of subschemas, for example, operation, man, equipment, produced material, consumed material, produced WIP, consumed WIP, and consumable information.

To make the real-time data of primitive event occurred at the distributed IoT devices meaningful manufacturing data which can be understood by the IoT-MS, an event model is used to establish the relationship mapping between the business context and the real-time field data in the manufacturing system. The event model has two main stages, namely event definition and executer. In the definition stage, the relationships such as the logic and sequence flows between the primitive events of the IoT/auto-ID devices and the meaningful manufacturing events will be defined and established. Then, in the executer stage, the result of the meaningful manufacturing events will be executed through the event executer according to the defined and established logic and sequence relationships of the relevant primitive events.

2. Sharing and exchanging module

Contrast to data processing module, sharing and exchanging module is simple. This module has two main components, namely information push and push list manager. The information push component is used to actively push the real-time information from the sharing and exchanging module to different users or heterogeneous information management systems. This component will be invoked once the primitive events have occurred at the distributed IoT/autoID devices. In order to share and integrate the real-time manufacturing data, the push list manager component is used to add or delete the information entry of different heterogeneous information management systems. Each information management system can define its personalized requirement, for example, the types of the real-time manufacturing information, remote transmit interfaces, and so on. Then, when information push component is invoked, the push list will be loaded at first and the standard real-time manufacturing information can be timely transmitted to the registered information management systems.

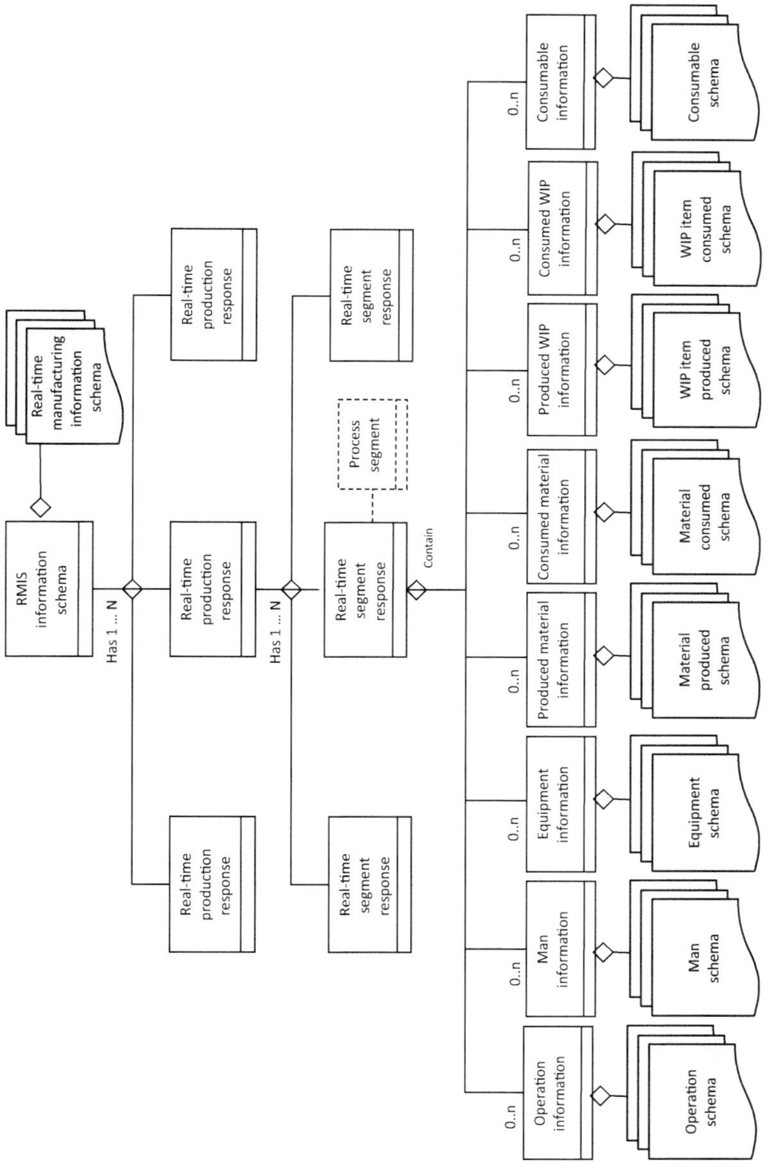

FIGURE 2.3 Real-time manufacturing information schema (RMIS).

2.5 THE WORKLOGIC OF IoT-MS

The worklogic of optimization of manufacturing systems using the IoT is described in Fig. 2.4, it has the following six steps.

1. IoT devices configuration
   The IoT devices configuration method is used to easily and effectively deploy the IoT devices. Here, the IoT devices include various types of sensors (e.g., RFID, temperature sensor, force sensor, and digital caliper) used to collect real-time data in manufacturing processes. Equipped with IoT devices, the real-time primitive event of production resources can be sensed and captured, creating a real-time and multisource manufacturing data sensing environment.
2. Smart model of bottom manufacturing resources
   Smart model of bottom manufacturing resources (e.g., machine tool and trolley) are constructed using IoT, computational intelligence, and information technology. Based on the model, manufacturing resources can actively interact and share information with each other in the manufacturing processes. The information also can be shared among the manufacturing resources and the upper management system.
3. Manufacturing resources configuration
   When the manufacturing tasks are obtained from manufacturing plant, manufacturing resources configuration module assign the tasks to the optimized

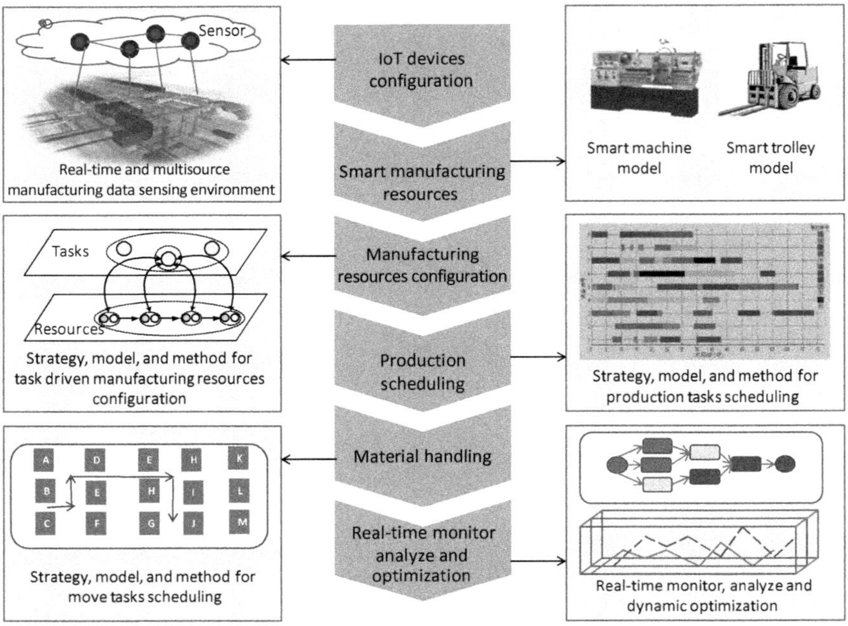

FIGURE 2.4 Worklogic of IoT-MS.

manufacturing resources according to the parameters and real-time status of each manufacturing resource.

4. Production task scheduling
Based on the results of manufacturing resources configuration, the production task scheduling module aims to sequence the tasks (in the process level) on the machines in order to minimize the makespan adopting intelligent algorithms. The results of the scheduling include the operation sequence as well as the start and end time of the tasks on the machine.
5. Material handling
By using IoT technology to material trolley, each trolley is an active entity which will request the transport tasks. An algorithm is designed to realize the dynamic optimization of the material handling tasks and transport route. Using the algorithm, transport tasks will be assigned to the optimal trolley according to the priority of tasks, maximum load and volume of the trolleys.
6. Real-time monitor, analyze, and optimization
During the manufacturing process, the production status of each process is perceived and fed back to the upper level management information system. According to the real-time status of the bottom manufacturing resources, the production performance of the plant manufacturing execution system is analyzed, in order to improve the transparency and traceability of the manufacturing system. Based on the analysis results, the tasks is reassigned to the manufacturing resources and rescheduled to realize real-time optimization.

2.6 DESCRIPTION OF THE CORE TECHNOLOGIES OF IoT-MS

Optimization of manufacturing systems using the IoT involves multidiscipline theories such as information technology, computer science, automation, industrial engineering, artificial intelligence, and so on. However, this book focuses mainly on the Internet of manufacturing things (resources), smart model of bottom manufacturing resources, and the real-time manufacturing data-driven actively sensing and dynamic optimization of the manufacturing system.

To implement the objective of the designed overall architecture of IoT-MS in Section 2.2, seven core technologies will be studied as shown in Fig. 2.5. The seven core technologies will be simply described here, which will be further discussed in details in the subsequent chapters.

1. Real-time and multisource manufacturing information perception system
The real-time and multisource manufacturing information perception is the basis of proactive sensing and dynamic scheduling of workshop production process. In order to achieve the information perception, various types of sensors need to be deployed and configured in a traditional manufacturing environment. For deployment and configuration of the sensor, there are three problems that need to be addressed. The first is how to achieve heterogeneous sensor device selection and configuration according to the types

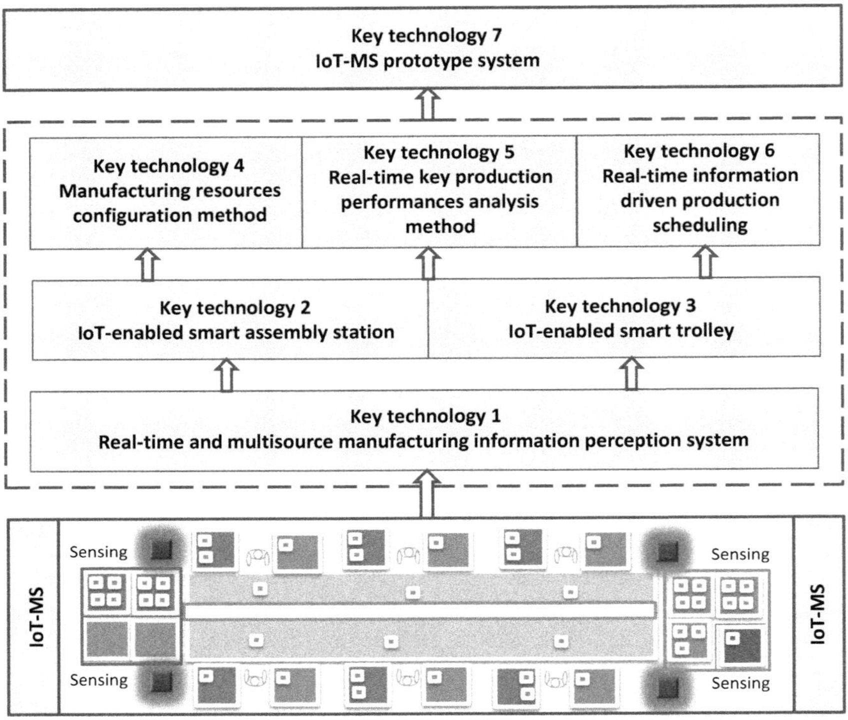

FIGURE 2.5 **Core technologies of IoT-MS.**

and amount of information collected. The second is how to achieve centralized management of the heterogeneous sensors. The third is how the raw data collected will be processed into understandable manufacturing information.

2. IoT-enabled smart assembly station
   By adopting automatic identification technologies, the assembly station is called as IoT-enabled smart assembly station, which can support real-time intelligent navigation. To provide optimal navigation for the assembly activities of each assembly station, three main services, namely the real-time assembly operating guidance service, collaborative production service among assembly stations, and real-time queuing service of task are designed. For the assembly line level, the disturbances and exceptions could be timely captured and reduced. The IoT-enabled smart assembly station will facilitate the real-time information-driven process monitoring and control between the line and stations.
3. IoT-enabled smart trolley
   Equipped with auto-ID devices, the trolleys can sense the different manufacturing resources (e.g., operator, material, pallet, location, and so on) and

actively request the move tasks, and the real-time status of each trolley could be timely tracked and traced. Based on these real-time and multisource data, intelligent navigation can be implemented to enhance the delivery efficiency. The problem is how to apply auto-ID devices and information technologies to enable the trolleys to have the capability of active sensing and intelligence so that the real-time material handling can be achieved.

4. CC-based manufacturing resources configuration method
The wide use of CC has prompted the development of CMfg. In CMfg, the distributed and heterogeneous manufacturing resources are virtualized and encapsulated into manufacturing cloud services, including manufacturing cell cloud service (MCCS) and manufacturing machine cloud service (MMCS). The manufacturing resources configuration method aims to select an optimal solution from large-scale service compositions (for both MCCSs and MMCSs),
5. Real-time key production performances analysis method
Timely and accurate perception and analysis of the production process key performance is critical to ensure the efficient and high quality operation of the manufacturing systems. Based on the real-time data, the key production performances analysis method is developed. The method can be divided into two parts. One is to map the relationship between the real-time key performances of complex production process (e.g., production order status, manufacturing cost, production quality and work in process inventory) and the related manufacturing resources. The other is to add value from the multisource manufacturing information based on decision trees, rule base, and data mining method. The analysis results of the real-time performance of manufacturing execution systems will be provided to different levels of the production managers.
6. Real-time information-driven production scheduling system
With the application of IoT technologies to manufacturing processes, real-time data has become more accessible and ubiquitous. Based on the real-time data, multiagent and auto-ID technologies are integrated to implement real-time shop-floor scheduling in a ubiquitous manufacturing environment. The multiagent real-time scheduling system includes four types of agents: machine agent, capability evaluation agent, real-time scheduling agent, and process monitor agent. These four agents are coordinated to realize the dynamic scheduling of production process of complex product, closing the loop of production planning and control.
7. IoT-MS prototype system
To demonstrate the feasibility of the proposed active perception and dynamic optimization framework, model, and technologies in manufacturing system, an IoT-enabled prototype system is designed, in order to improve the transparency of the manufacturing process. In addition, the real-time abnormal event detection in production process can ensure the high efficiency and quality production of manufacturing tasks.

RFID systems are deployed to collect the real-time production data, making the operational status of the shop floor become transparent. Based on data analysis, the abnormal information of the key events in production is diagnosed, and the production performances are evaluated, so as to realize the dynamic optimization of the production process. Using the prototype system, employees including machine operators and distribution workers, are guided by the visual interface of the devices to perform tasks. This can enhance the overall plant efficiency. In addition, based on the historical production information, managers can make decisions about future production tasks.

REFERENCES

[1] K. Ashton, That 'Internet of Things' thing, RFiD J. 22 (7) (2009) 97–114.
[2] D. Herrero-Pérez, H. Martínez-Barberá, Modeling distributed transportation systems composed of flexible automated guided vehicles in flexible manufacturing systems, IEEE Transac. Ind. Inform. 6 (2) (2010) 166–180.
[3] M.P. Groover, Automation, production systems, and computer integrated manufacturing, Prentice Hall Press, (2007).
[4] L. Sanchez, R. Nagi, A review of agile manufacturing systems, Int. J. Prod. Res. 39 (16) (2001) 3561–3600.
[5] C.A. Rusinko, Green manufacturing: an evaluation of environmentally sustainable manufacturing practices and their impact on competitive outcomes, IEEE Trans. Eng. Manage. 54 (3) (2007) 445–454.
[6] F. Jovane, H. Yoshikawa, L. Alting, C.R. Boër, E. Westkamper, D. Williams, M. Tseng, G. Seliger, A.M. Paci, The incoming global technological and industrial revolution towards competitive sustainable manufacturing, CIRP Ann. 57 (2) (2008) 641–659.
[7] F. Tao, D. Zhao, Y. Hu, Z. Zhou, Resource service composition and its optimal-selection based on particle swarm optimization in manufacturing grid system, IEEE Trans. Ind. Inform. 4 (4) (2008) 315–327.
[8] H. Meier, R. Roy, G. Seliger, Industrial product-service systems—IPS2, CIRP Ann. 59 (2) (2010) 607–627.
[9] Y. Zhang, G. Zhang, J. Wang, S. Sun, S. Si, T. Yang, Real-time information capturing and integration framework of the Internet of manufacturing things, Int. J. Comput. Integr. Manuf. 3052 (December) (2014) 1–12.
[10] D. Tranfield, D. Denyer, P. Smart, Towards a methodology for developing evidence-informed management knowledge by means of systematic review, Br. J. Manage. 14 (2003) 207–222.
[11] E. Auschitzky, M. Hammer, A. Rajagopaul, How big data can improve manufacturing, McKinsey Co., no. July, 2014, pp. 1–4.
[12] J. Gubbi, R. Buyya, S. Marusic, M. Palaniswami, Internet of Things (IoT): a vision, architectural elements, and future directions, Futur. Gener. Comput. Syst. 29 (7) (2013) 1645–1660.
[13] F.J. Villanueva, D. Villa, F. Moya, M.J. Santofimia, J.C.L. Pez, Internet of things architecture for an RFID-based product tracking business model, in: Proceedings of the Sixth International Conference on Innovative Mobile and Internet Services in Ubiquitous Computing, IMIS, 2012, pp. 811–816.
[14] H. Cai, L. Xu, B. Xu, C. Xie, S. Qin, L. Jiang, IOT-based configurable information service platform for product lifecycle management, IEEE Trans. Ind. Inform. 2 (2014) 1.

[15] V. Noresson, N.G. Ohlson, M. Nilsson, Grocery Manufacturers of America, Washington, DC, (2002).
[16] P. Hank, S. Muller, O. Vermesan, J. Van Den Keybus, Automotive Ethernet: in-vehicle networking and smart mobility, in: Proceedings of the Conference on Design, Automation and Test in Europe, 2013, pp. 1735–1739.
[17] E. Polycarpou, L. Lambrinos, E. Protopapadakis, Smart parking solutions for urban areas, in: 2013 IEEE 14th International Symposium on a World of Wireless, Mobile and Multimedia Networks, WoWMoM, 2013.
[18] G. Amato, F. Falchi, F. Rabitti, Landmark recognition in VISITO Tuscany, Commun. Comput. Inf. Sci. 247 (2012) 1–13.
[19] E. Ancillotti, R. Bruno, M. Conti, Smoothing peak demands through aggregate control of background electrical loads, in: 2014 IEEE PES Innovative Smart Grid Technologies Conference, ISGT, 2014, pp. 1–5.
[20] J. Lu, T. Sookoor, V. Srinivasan, G. Gao, B. Holben, J. Stankovic, E. Field, K. Whitehouse, The smart thermostat: using occupancy sensors to save energy in homes, Proc. ACM SenSys 55 (2010) 211–224.
[21] E. Sun, X. Zhang, Z. Li, The Internet of things (IOT) and cloud computing (CC) based tailings dam monitoring and pre-alarm system in mines, Saf. Sci. 50 (4) (2012) 811–815.
[22] F. Delmastro, Pervasive communications in healthcare, Comput. Commun. 35 (11) (2012) 1284–1295.
[23] F. Tao, L. Zhang, V.C. Venkatesh, Y. Luo, Y. Cheng, Cloud manufacturing: a computing and service-oriented manufacturing mode, Proc. Inst. Mech. Eng. B 225 (2011) 1969–1976.
[24] A. Iosup, S. Ostermann, N. Yigitbasi, R. Prodan, T. Fahringer, D. Epema, Performance analysis of cloud computing services for many-tasks scientific computing, IEEE Trans. Parallel Distrib. Syst. 22 (6) (2011) 931–945.
[25] F. Tao, L. Zhang, K. Lu, D. Zhao, Research on manufacturing grid resource service optimal-selection and composition framework, Enterp. Inf. Syst. 6 (2) (2012) 237–264.
[26] B.H. Li, L. Zhang, L. Ren, X.D. Chai, F. Tao, Y.L. Luo, Y.Z. Wang, C. Yin, G. Huang, X.P. Zhao, Further discussion on cloud manufacturing, Jisuanji Jicheng Zhizao Xitong (Comput. Integr. Manuf. Syst. CIMS) 17 (3) (2011) 449–457.
[27] W. Jiang, J. Ma, X. Zhang, H. Xie, Research on cloud manufacturing resource integrating service modeling based on cloud-Agent, in: ICSESS 2012—Proceedings of 2012 IEEE Third International Conference on Software Engineering and Service Science, 2012, pp. 395–398.
[28] Y. Bao, L. Ren, L. Zhang, X. Zhang, Y. Luo, Massive sensor data management framework in cloud manufacturing based on Hadoop, in: IEEE International Conference on Industrial Informatics (INDIN), 2012, pp. 397–401.
[29] X. Xu, From cloud computing to cloud manufacturing, Robot. Comput. Integr. Manuf. 28 (1) (2012) 75–86.
[30] D. Wu, D.W. Rosen, L. Wang, D. Schaefer, Cloud-based design and manufacturing: a new paradigm in digital manufacturing and design innovation, Comput. Des. 59 (2015) 1-L14.
[31] D. Wu, J.L. Thames, D.W. Rosen, D. Schaefer, Towards a cloud-based design and manufacturing paradigm: Looking backward, looking forward, in: Proceedings of the ASME Design Engineering Technical Conference, vol. 2, no. Parts A and B, 2012, pp. 315–328.
[32] A.L.K. Yip, J.R. Corney, A.P. Jagadeesan, Y. Qin, A product configurator for cloud manufacturing, in: Volume 2 Systems; Micro and Nano Technologies; Sustainable Manufacturing, vol. 2, 2013, p. V002T02A010.
[33] L. Wang, Machine availability monitoring and machining process planning towards Cloud manufacturing, CIRP J. Manuf. Sci. Technol. 6 (4) (2013) 263–273.

[34] F. Tao, L. Zhang, Y. Hu, Resource Service Management in Manufacturing Grid System. 2012. Scrivener Publisher.
[35] Y. Lu, X. Xu, J. Xu, Development of a Hybrid Manufacturing Cloud, J. Manuf. Syst. 33 (4) (2014) 551–566.
[36] X. Yue, H. Cai, H. Yan, C. Zou, K. Zhou, Cloud-assisted industrial cyber-physical systems: an insight, Microprocess. Microsyst. 39 (8) (2015) 1262–1270.
[37] D. Mourtzis, M. Doukas, C. Lalas, N. Papakostas, Cloud-based integrated shop-floor planning and control of manufacturing operations for mass customisation, Procedia CIRP 33 (2015) 9–16.
[38] R. Gao, L. Wang, R. Teti, D. Dornfeld, S. Kumara, M. Mori, M. Helu, Cloud-enabled prognosis for manufacturing, CIRP Ann. 64 (2) (2015) 749–772.
[39] N. Liu, X. Li, A resource virtualization mechanism for cloud manufacturing systems, in: M. van Sinderen, P. Johnson, X. Xu, G. Doumeingts (Eds.), Proceedings of the Enterprise Interoperability: Fourth International IFIP Working Conference, IWEI 2012, Harbin, China, September 6–7, 2012, Berlin, Heidelberg, Springer, 2012, pp. 46–59.
[40] J. Gao, Y. Yao, V.C.Y. Zhu, L. Sun, L. Lin, Service-oriented manufacturing: a new product pattern and manufacturing paradigm, J. Intell. Manuf. 22 (3) (2011) 435–446.
[41] X. Geng, X. Chu, D. Xue, Z. Zhang, An integrated approach for rating engineering characteristics' final importance in product-service system development, Comput. Ind. Eng. 59 (4) (2010) 585–594.
[42] M.-J. Kang, R. Wimmer, Product service systems as advanced system solutions for sustainability, in: EKC2008 Proceedings of the EU–Korea Conference on Science and Technology, 2008, vol. 124, pp. 191–199.
[43] M.J. Goedkoop, C.J.G. Van Halen, H.R.M. Te Riele, P.J.M. Rommens, Product Service systems, Ecological and Economic Basics, vol. 36, no. March, 1999.
[44] O.K. Mont, Clarifying the concept of product-service system, J. Clean. Prod. 10 (3) (2002) 237–245.
[45] X. Yang, P. Moore, J.S. Pu, C.B. Wong, A practical methodology for realizing product service systems for consumer products, Comput. Ind. Eng. 56 (1) (2009) 224–235.
[46] G.Q. Huang, T. Qu, Y. Zhang, H.D. Yang, RFID-enabled product-service system for automotive part and accessory manufacturing alliances, Int. J. Prod. Res. 50 (14) (2012) 3821–3840.
[47] N. Maussang, P. Zwolinski, D. Brissaud, Product-service system design methodology: from the PSS architecture design to the products specifications, J. Eng. Des. 20 (4) (2009) 349–366.
[48] H.J. Long, L.Y. Wang, J. Shen, M.X. Wu, Z. Jiang, Product service system configuration based on support vector machine considering customer perception, Int. J. Prod. Res. 51 (18) (2013) 5450–5468.
[49] B. Yoon, S. Kim, J. Rhee, An evaluation method for designing a new product-service system, Expert Syst. Appl. 39 (3) (2012) 3100–3108.
[50] B. Laperche, F. Picard, Environmental constraints, Product-Service Systems development and impacts on innovation management: learning from manufacturing firms in the French context, J. Clean. Prod. 53 (2013) 118–128.
[51] Y. Zhang, T. Qu, O. Ho, G.Q. Huang, Real-time work-in-progress management for smart object-enabled ubiquitous shop-floor environment, Int. J. Comput. Integr. Manuf. 24 (5) (2011) 431–445.
[52] S. Energy, Automation, Inc. 2006.'Why integrate MES and ERP? Because you can't afford not to,' Siemens Whitepaper, pp. 1–8.
[53] H. Panetto, A. Molina, Enterprise integration and interoperability in manufacturing systems: trends and issues, Comput. Ind. 59 (7) (2008) 641–646.
[54] J.M. Rodriguez, M. Crasso, A. Zunino, M. Campo, Improving Web service descriptions for effective service discovery, Sci. Comput. Program. 75 (11) (2010) 1001–1021.

[55] X. Kong, C. Lin, Y. Jiang, W. Yan, X. Chu, Efficient dynamic task scheduling in virtualized data centers with fuzzy prediction, J. Netw. Comput. Appl. 34 (4) (2011) 1068–1077.
[56] ANSI/ISA–95, Instrumentation, systems, and automation, Part 1 enterprise-control system integration, 2000.
[57] L. Gould, B2MML explained, Automotive Design & Production, 2007, www.autofieldguide.com/articles/b2mml-explained.
[58] V. Anaya, G. Berio, M. Harzallah, P. Heymans, R. Matuleviius, A.L. Opdahl, H. Panetto, M.J. Verdecho, The Unified Enterprise Modelling Language-Overview and further work, Comput. Ind. 61 (2) (2010) 99–111.
[59] Y. Zhang, G.Q. Huang, T. Qu, O. Ho, S. Sun, Agent-based smart objects management system for real-time ubiquitous manufacturing, Robot. Comput. Integr. Manuf. 27 (3) (2011) 538–549.
[60] P. Vichare, A. Nassehi, S. Kumar, S.T. Newman, A Unified Manufacturing Resource Model for representing CNC machining systems, Robot. Comput. Integr. Manuf. 25 (6) (2009) 999–1007.
[61] S.Y. Shi, R. Mo, H.-C. Yang, Z.-Y. Chang, Z.-F. Chen, An implementation of modelling resource in a manufacturing grid for resource sharing, Int. J. Comput. Integr. Manuf. 20 (2–3) (2007) 169–177.
[62] N. Liu, X. Li, W. Shen, Multi-granularity resource virtualization and sharing strategies in cloud manufacturing, J. Netw. Comput. Appl. 46 (2014) 72–82.
[63] F. Tao, Y. Laili, L. Xu, L. Zhang, FC-PACO-RM: a parallel method for service composition optimal-selection in cloud manufacturing system, IEEE Trans. Ind. Inform. 9 (4) (2013) 2023–2033.
[64] N. Do, Developing a product data model using the STEP PDM schema for integrating specifications of technical services, Int. J. Comput. Integr. Manuf. 27 (4) (2013) 382–399.
[65] L. Yan, Z.M. Ma, Extending engineering data model for web-based fuzzy information modeling, Integr. Comput. Aided Eng. 20 (4) (2013) 407–420.
[66] H.M. Shih, Migrating product structure bill of materials Excel files to STEP PDM implementation, Int. J. Inf. Manage. 34 (4) (2014) 489–516.

Chapter 3

Real-Time and Multisource Manufacturing Information Sensing System

3.1 INTRODUCTION

The recent developments in wireless technologies [e.g., radio frequency identification (RFID), auto-ID and smart meters] have created a new era of the Internet of things (IoT) [1–4]. IoT represents uniquely distinguishable objects and their virtual representations in an Internet-alike structure. With the support and application of IoT technologies, the potentially intelligent and real-time operators of 4Cs (i.e., sensing and Connection, Communication, Computing, and Control) to both physical and virtual objects can be realized [5,6]. Furthermore, by integrating IoT with production, logistics, and services in the current industrial fields, it is possible to transform today's factory into the smart factory with significant economic potential [7–9]. The smart factory refers to a environment-sensitive manufacturing surroundings that can handle disturbances in real-time production using decentralized information and communication structures for an optimal management of manufacturing processes [10–13].

In traditional manufacturing shop floor, the information is manually collected, which is often laggard, inaccurate, and discrete. Thus, it is difficult for users to find exceptional events timely. When the exception happens, it will spread gradually and further aggravates the turbulences of manufacturing process. Therefore, it is necessary for manufacturers to improve their real-time and multisource manufacturing information (RMMI) sensing ability with the advanced IoT technologies and models to increase the production efficiency.

To meet the requirements of real-time information sensing, the IoT technologies have been adopted to shop floor for capturing and sharing the real-time and multisource manufacturing information. Huang et al. proposed a conceptual wireless manufacturing (WM) framework [14]. Zhang et al. presented a real-time information capturing and integration architecture of the Internet of manufacturing things [2]. Zhang et al. presented the idea of cloud manufacturing (CMfg), and the architecture, characteristics, and core enabling technologies are extensively researched [15]. Tao et al. provided an IoT and bill of material

Optimization of Manufacturing Systems Using the Internet of Things
http://dx.doi.org/10.1016/B978-0-12-809910-0.00003-2

(BOM)-based life cycle assessment of energy-saving and eRMMISSion-reduction of products [5]. Fang et al. discussed an agent-based gateway operating system (GOS) for the RFID-enabled ubiquitous manufacturing enterprise [16].

Despite the significant progresses, the following challenges still exist in many manufacturing companies. The first challenge is how to apply the advanced IoT technologies to traditional manufacturing resources, thus the status of resources could be timely sensed and reflected during production processes. The second challenge is how to use a standard model to manage the multi-types of smart sensors, so that they can be managed in a "plug-and-play (PnP)" fashion, and can be easily reused and reconfigured for different manufacturing purpose without professional knowledge. The third challenge is how to easily and effectively process and share the RMMI, so that the seamless and dual-way connectivity and interoperability between different users and shop-floor front-line can be realized.

Considering the advantages of IoT technologies and the requirements of RMMI sensing, this article discusses an overall real-time and multisource manufacturing information sensing system (RMMISS). The presented RMMISS aims to eliminate the information gap between enterprise application services and discrete production resources such as trolleys, machines, and so on. It will manage the various sensors in a PnP pattern and deliver the RMMI reliably and dynamically.

The rest of the chapter is organized as follows. After the works related to this research is reviewed in Section 3.2, Section 3.3 presents an overview of real-time and multisource RMMISS and briefly introduces its key technologies. Section 3.4 gives the key technologies of configuration of multiple sensors while Section 3.5 presents the key technologies of sensor manager. Section 3.6 presents key technologies of multisource manufacturing information processing and sharing. Finally, a case is presented in Section 3.7.

3.2 RELATED WORKS

Three streams of related works are relevant to this research. These include (1) real-time manufacturing data capturing, (2) sensor management, and (3) manufacturing information processing and sharing.

3.2.1 Real-Time Manufacturing Data Capturing

Due to the rapid development in IoT technologies, real-time manufacturing data capturing attracts increasing attention from both industrial field and academia. RFID is the key technology of IoT, and it has been widely studied to capture the manufacturing data. To tackle scheduling inefficiency resulting from paper-based identification and manual data collections, Zhong et al. presented an RFID-enabled real-time manufacturing execution system, which is fulfilled by systematically configuring RFID devices on the shop floor to track and trace manufacturing objects and collect real-time production data [17]. Yang et al. presented an RFID-enabled indoor locating method for a real-time manufacturing execution system using online sequential extreme learning machine [18]. Lee et al. proposed an RFID-based resource assignment system for garment manufacturing [19]. Liu et al. designed a new production management system by integrating with an RFID-enabled real-time data capturing system for Loncin motorcycle assembly line [20]. The RFID-based remote monitoring system is proposed to provide a transparent and visible information flow for supply chain and enterprise internal resource management [21].

At the same time, other auto-ID devices are also used to sense the real-time status of manufacturing resources. For example, the concept of smart objects was proposed by Zhang et al. to represent the behaviors of the heterogeneous auto-ID devices for capturing real-time manufacturing data [22]. By using IoT, Qu et al. researched the demand of dynamic production logistics synchronization (PLS) for a manufacturer. Advanced cloud manufacturing (CMfg) and IoT technologies were systematically combined to support a smart PLS control method with multilevel dynamic flexibility [23]. Jiang et al. presented a material flow management model based on "event-triggering time-state" graphical schema [24]. Yew et al. described an amplified reality manufacturing system that targets to greatly increase the information sensing ability of different kinds of workers in a manufacturing equipment and the workers can interact with the surroundings [25]. Amit and Pieper proposed a noninvasive methodology and development of a software application to monitor real-time machine status, energy usage, and other machining parameters for a legacy machine tool using power signal analysis [26]. Mori et al. proposed a system for machine tool manufacturers to monitor and maintain their customers' machine tools remotely, three key technologies, that is, the communication modes, information schemas and data processing manners, are detailed discussed in the paper [27]. Shen et al. proposed a concept called iShopFloor, that is, an intelligent shop floor based on the Internet, web, and agent technologies, and it provided the framework for components of a complex control system to work together as a whole rather than as a disjoint set [28].

3.2.2 Sensor Management

The sensor management is essential in providing reliable manufacturing data. However, the current connect number, sampling rate, and signal types of sensors are generally restricted by the device, a unified sensor management model plays an important role in manufacturing information sensing. Chi et al. proposed a reconfigurable smart sensor interface for industrial wireless sensor network in IoT environment, in which complex programmable logic device was used as the central manager [29]. Mayer et al. presented a service composition system that enables the goal-driven configuration of smart environments for end users by combining semantic metadata and reasoning with a visual modeling tool [30]. Boonma and Suzuki proposed a biologically inspired middleware architecture for self-managing wireless sensor networks [31]. Jung et al. presented a new architecture of Hadoop-based distributed sensor node management system for distributed sensor node management using Hadoop map reduce framework and distributed file system [32]. Wang et al. proposed a Hadoop-based approach for web service management in telecommunication and Internet domains, the basic idea is to adopt two components of Hadoop, that is, HBase, and MapReduce, to manage web services [33]. Kulvatunyou et al. analyzed the OWL-based semantic mediation approaches to enhance manufacturing service capability models [34]. Fröhlich and Wanner presented a run-time support environment for wireless sensor network applications based on the operating system. It allowed the applications to configure their communication channel according to their needs and acquire sensor data through a unified application program interface (API) [35]. Hu et al. proposed a real-time discrete event–based monitoring system for RFID-enabled shop-floor monitoring in manufacturing enterprises. The system used rigorous mathematical techniques for event construction, state prediction, and disturbance detection that are appropriate for big-data environments of current intricate manufacturing systems [36]. Fang et al. presented an agent-based GOS to manage the RFID sensors for the ubiquitous company [16]. Considering the requirements of integration between production activities and enterprise information systems, an integration framework for RFID middleware based on business process rule and data stream technologies are presented by Li and Li [37].

3.2.3 Manufacturing Information Processing and Sharing

Manufacturing information captured by multiple sensors are often with different format, thus they need to be processed to obtain information in standard schema, and how to share the information with different EISs are also an important question. To address this demand, Anaya et al. described a unified enterprise modeling language, which aimed to backup integrated use of enterprise information models expressed using different languages [38]. Pantazoglou et al. proposed Proteus, a generic query model for the discovery of operations offered by heterogeneous services [39]. Al-Fuqaha et al. designed a rule-based intelligent gateway that can bridge the gap between existing IoT protocols to manage the efficient integration of parallel IoT services [40]. In order to obtain the formal description of manufacturing capability, Luo et al. studied a modeling and description method of multidimensional information for manufacturing capability in CM system [41]. Zhang et al. presented a services encapsulation and virtualization access model for manufacturing machine by combing the IoT techniques and cloud computing [42], and a task-driven manufacturing cloud service proactive discovery and optimal configuration method [43]. Russom et al. discussed the implementation of a configurable laboratory information management system for use in cellular process development and manufacturing [44]. To help engineers use digitization practically and efficiently, Kojima et al. proposed a method based on manufacturing case data that had a direct relation to manufacturing operations, where the data were represented in XML schema [45]. Pang et al. discussed the data-source interoperability service for heterogeneous information integration in ubiquitous enterprises [46]. Chungoora et al. studied the interoperable manufacturing knowledge systems concept, which used an expressive ontological method that derived the improved configuration of product life cycle management systems for manufacturing knowledge sharing [47]. Wu et al. developed a method to improve the integration and sharing of product knowledge during the development phase, the study established product development knowledge integration ontology [48]. Dhokia et al. developed a data model and a prototype information sharing platform for DEMAT machine tools, and the platform provided a unique method to monitor the life cycle status of machines [49]. Qiu et al. proposed an IoT-enabled Supply Hub in Industrial Park for improving the effectiveness and efficiency of sharing physical assets and services [50].

3.3 OVERALL ARCHITECTURE OF REAL-TIME AND MULTISOURCE RMMISS

This research focuses mainly on the real-time and multisource manufacturing information sensing in discrete manufacturing field. The overall architecture of the proposed RMMISS is shown in Fig. 3.1. Mainly it consists of four modules, namely deployment of multiple sensors, multisensor manager, multisource manufacturing information processing, and sharing and application services. They are described as follows.

3.3.1 Deployment of Multiple Sensors

This module aims to construct a smart environment by extending the advanced IoT technologies to traditional manufacturing field. First, the information requirements of the manufacturing system are analyzed. Second, the sensors types (e.g., RFID, barcode, smart meters, and so on) are selected according to the information requirements. As a result, the sensors are deployed in the shop-floor frontline, and the RMMI can be timely sensed by the up-level applications. More details will be presented in Section 3.4.

3.3.2 Multiple Sensors Manager

As shown in the middle of Fig. 3.1, multiple sensors manager is used to manage the sensors, so that the sensors can work properly. First, the sensors are registered into the multisensor manager, where the parameters of the sensors are set. Second, the sensor drivers are installed to ensure the sensors operate reliably. Third, the binding models between the resources and sensors are established. At last, the operating status of sensors is monitored by this module, so that the sensor exceptions can be eliminated as soon as possible. More details will be presented in Section 3.5.

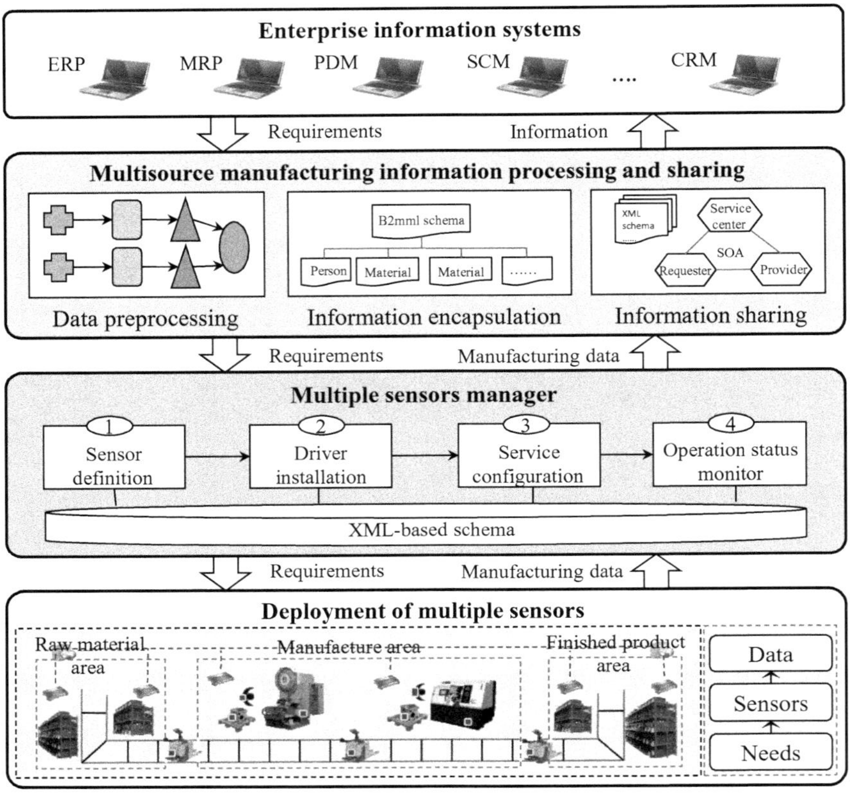

FIGURE 3.1 **Overall architecture of real-time and multisource RMMISS.**

3.3.3 Multisource Manufacturing Information Processing and Sharing

Huge amounts of real-time manufacturing data are captured by the sensors. However, most of the data are duplicated, meaningless, and unreliable, and only limited quantities of manufacturing events are taken care by the up-level users. The multisource manufacturing information processing and sharing module is used to meet the gaps between the raw sensor data and EISs requirements. Three kinds of contents are discussed in this module, namely data preprocessing, information encapsulation, and information sharing mechanism. Data preprocessing is used to aggregate the discrete data into resource level event. Information encapsulation is used to encapsulate into a standard information template which is based on B2MML schema. Information sharing mechanism is used to transfer the qualified data into the related users. More details will be discussed in Section 3.6.

3.4 DEPLOYMENT OF MULTISENSORS

3.4.1 Description of Multisource Manufacturing Information

Since many manufacturing resources are involved in the shop floor and their status and capacity data are changing continuously, a vast number of manufacturing data are created. Thus, it is important to select the key monitor points of manufacturing information. As shown in Fig. 3.2, five kinds of key information are interested by the managers, that is, machine, process quality, manufacturing objects, worker, and environment, and they are described as follows.

1. Manufacturing information related to machine
The status of machine plays an important role in improving the economic returns. If the machine works properly, the quantity and quality of production output can be guaranteed. Otherwise, the production may be suspended, which will lead to short supply of work in progress (WIP) or product. In real manufacturing process, the critical units and weak link of the machines are always the critical points for inspection, for example, the transmission system, the machining section, the servo system, and the supporting component. In general, there is a period of time before the failure event happens, during which the failure can be reflected by the data of sound, heat, and variation, it will be helpful to monitor the real-time data of machine in the reliability analysis process of the machine.

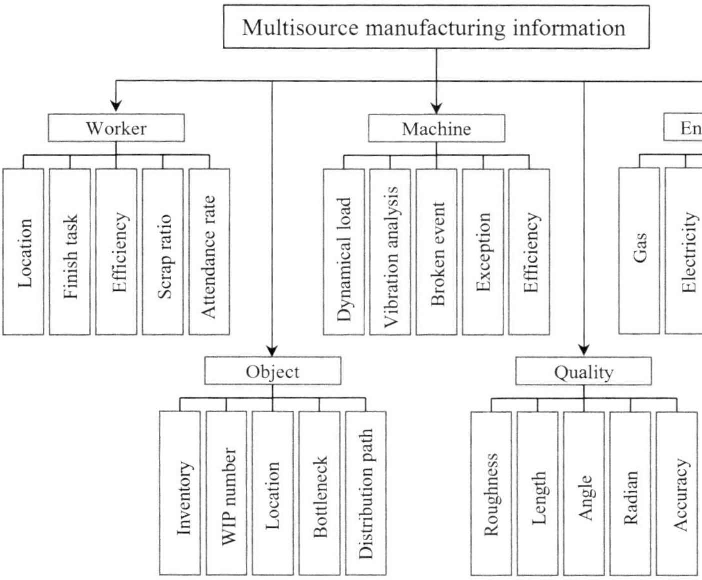

FIGURE 3.2 Multisource manufacturing information.

2. Manufacturing information related to process quality
The process quality of WIP and production plays an important role in the usage stage of production. The key indicator of process quality includes two sides: processing precision and surface quality. Processing precision is defined as closeness between physical parameters and technological parameters, and it can be categorized as dimensional accuracy, shape accuracy, or positional accuracy. Surface quality is the detection quality of production surface, that is, the surface roughness and mechanical properties.
3. Manufacturing data related to manufacturing objects
The manufacturing objects include the raw materials, the WIP, and the finished products. The data of raw materials, for example, the location, number, and status, can give useful advises for the inventory replenishment. The status of WIPs (e.g., the real-time progress, the manufacturing machine, waiting time, and so on) can reflect the production bottleneck and ensure the WIP flow follows the right direction. The data of finished products (e.g., the number, scrap ratio, and so on) display the production capacity of the shop floor.
4. Manufacturing data related to worker
The manufacturing data related to worker consists of the personnel assigned, skill, seniority, manufacturing record, etc. These data can provide the essential reference for the staff assignment. Since the machine failure cannot be avoided easily, the effective management of worker and manufacturing objects is potential in improving the production efficiency and ensuring that the shop floor obtains the largest production capacity.
5. Manufacturing data related to environment
The environment is another important factor in shop floor, and the excellent environment can stimulate the maximum ability of person and ensure the machine work properly. On one hand, good environment can reduce fatigue and stress for the worker; on the other hand, with the wide use of the intelligent machine, it is necessary to keep a strict environment for the normal production. Otherwise, the abnormal production activities will happen, for example, the dust may influence the manufacturing accuracy, the high temperature or humidity may affect the manufacturing ability, and so on.

3.4.2 Multiple Sensors Selection

As discussed previously, the information requirements in manufacturing field involves many things in productive processes, such as materials, WIPs, personnel, tools, equipment, vehicles, and so on. The real-time condition monitor of the manufacturing things has large impact on the performance of the entire production system. According to our recent investigation of several collaborative manufacturing enterprises, many of them use the manual system for data collection. Thus, the collected data are often laggard in time, prone to errors, and tedious. It is a daunting task to trace and track WIP items in a large manufacturing company. Besides, manual identification sheets are frequently damaged, lost, or misplaced, and shop-floor operators are busy with operations that are supposed to add values to products. As a result, the captured information can not accurately and promptly reflect the real-time status. In order to capture the real-time information of manufacturing resources, we choose some typical sensor and deploy them on suitable objects. The deployment information is shown in Table 3.1.

3.5 MULTIPLE SENSORS MANAGER

After deployment of multiple sensors, the multiple sensors manager will manage the multiple sensors dynamically to ensure they work properly.

First, when new sensors are deployed, the sensors are registered into the system, and the inherent parameters of the sensors are defined. The parameters include the type of the sensor, the frequency (UHF or HF), interface (USB or COM), connection port, production information, etc.

As heterogeneous sensors have different drivers, relevant software, communication mode, and interface, it's hard to drive each sensor after it connects to the RMMISS. To solve this problem, two kinds of sensor-driven mechanism are used, that is, the standard interface and driver library. After the sensor is registered into the system, the standard interfaces are used to drive the sensor. If the sensor cannot be driven successfully, the system will download the third-party driver from the Internet according to sensor type, brand, and version, and then install it on the system and update the driver library with the latest edition.

Since every sensor has its own work pattern, it is difficult to control these sensors in a uniform mode under the same platform. Thus, service-oriented architecture (SOA) is adopted in the multiple sensors manager, and the heterogeneous sensors can be published, searched, and invoked through the Internet. The function of each sensor will be wrapped as standard web service first, where each web service will get a single service address and service ID. As a result, the sensors can be managed in a PnP pattern, and the multisource and heterogeneous manufacturing data can be captured easily.

3.6 MULTISOURCE MANUFACTURING INFORMATION CAPTURING AND SHARING

Multisource manufacturing information capturing and sharing module is composed of three submodules, namely, data preprocessing, information encapsulation, and information sharing.

3.6.1 Data Preprocessing

After the multiple sensors are deployed and configured into the traditional manufacturing field, the real-time data of manufacturing resources will be captured continually. However, only a few data reflects the useful manufacturing information.

TABLE 3.1 Sensor Deployment in Manufacturing Shop Floor

Type	Hardware interface	Target manufacturing resource	Function
RFID reader and tag	RS232	Critical component, tool and critical port, etc.	Track the objects attached with RFID tags when they close the reader
Barcode reader and barcode	Keyboard interface	Pallet, material, and product	Obtain the information through barcode
Digital caliper	USB	Parts, subfinished or finished products	Measure the length, width, depth, inner and outer diameter in a high accuracy requirement
Temperature and humidity sensor	RJ45	Room, workshop, and warehouse	Obtain the temperature and humidity data
Displacement sensor	RS232/ RS485	Machine	Detect the rotor's state of motion in rotary machine, for example, rotor's radial vibration, axial vibration, rotate speed, shaft centerline orbit, the position of axis, etc.
Acceleration sensor	Serial	Machine	Measure the acceleration of vibration occurring machine. Find the cause of mechanical vibration and mechanical noise
RFID handheld reader and tag	Wi-Fi	Materials, product, pallet, etc.	Track the objects where the fixed reader cannot be used
Smart electricity meters	RS485	Workshop and energy-consuming machine	Capture the parameters of electricity consumption, such as the power rate, total consumption of electricity
Smart water meters	RS485	Workshop and water-consuming machine	Capture the parameters of water consumption, total consumption of water

To fulfill this gap, a two-level event model is used to process the raw sensor data into resource status data. Manufacturing data preprocessing helps users to obtain more meaningful and actionable resources information from the large amount of raw sensor events.

Primitive events (PEs) are events captured by the IoT technologies, which can be obtained in a large volume due to the reading characteristic of high-speed and automatic reading. As the data are often missed, duplicated, and inaccurate, a preprocessing method is necessary to provide data with worthy quality.

> *Definition 3.1*: Primitive events can be represented as $PE = (S,O,T)$, where S denotes the unique sensor ID, O denotes the unique object ID, while T denotes the sensing time.

Basic events (BEs) are resource level events, which are formed by the aggregation of qualified primitive events, and the BEs reflect the real-time space or space change of one or one class of manufacturing resources. The BEs constitute the lowest meaningful events that are taken care by the upper applications, five types of BEs are considered in this chapter: material distribution events (BE1), WIP circulating events (BE2), processing and assembly events (BE3), quality detection events (BE4), and storage events (BE5).

> *Definition 3.2*: BEs can be represented as $BE_{i,j}^{k} = (e/es, loc, context, T)$, $j = (1,2,3,4,5)$, $i = (1\cdots m)$, $k = (1\cdots n)$.
> Here $BE_{i,j}^{k}$ denotes jth kind of event at location i for object k, m, and n are the total number of the sensors and objects. E or es present the EPC of object; loc gives the position where event occurs. $Context$ is used to interpret the event attributes, such as the process quality, the circulating executer, the manufacturing machine, and so on. T denotes the time, and it can be either a time point or an interval of time.

To model and obtain the BEs efficiently, the timed transition color Petri nets (TTCPNs) are used to analysis the primitive events, which will be discussed in detail in Chapter 7. Referring to the quality detection process in Fig. 3.3A, the TTCPN is built up as in Fig. 3.3B. In the graphical representation, places are presented as circles, transitions are presented as bars, tokens are presented by colored dots, and the input and output are presented as arcs which are connected by arrows.

When a WIP enters into the quality detection process, it needs to wait in the in-buffer, and the place "WIP Arrived" obtains a token to record the WIP enter event. If the person is ready, the WIP will be initially detected. The person status is shown in place "person," while the initial detection process is represented

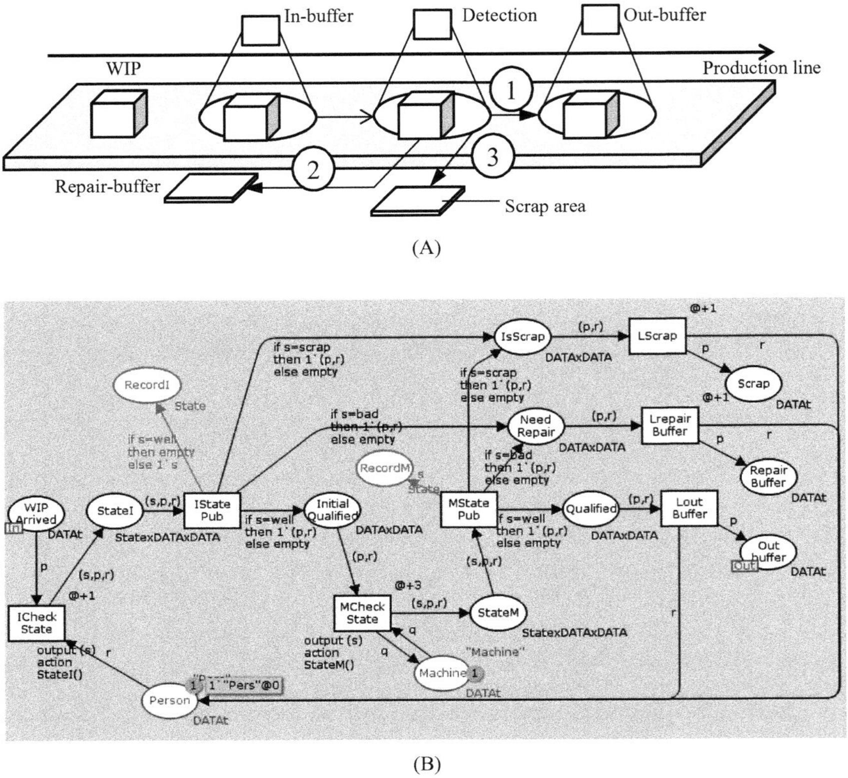

FIGURE 3.3 TTCPN model for quality detection process. (A) Case description. (B) TTCPN model.

by the state of transition "ICheck," and the result of initial check is denoted by the place "StateI." After the state is initially checked, the checked result will be published into the system, and the WIP is distributed into the relative area. Three places, that is, IsScrap, NeedRepair and Initial Qualified, are considered in the system to denote the three status of WIP, respectively. Then, the WIP is further detected by the machine. Based on the detection result, the WIP will be distributed into different areas, which are denoted as the place "Scrap," "Repair Buffer," and "Out-buffer," respectively.

3.6.2 Information Encapsulation

Since the output of each sensor is isolated manufacturing data, it is difficult to share them with heterogeneous EISs. A business to manufacturing markup language (B2MML)-based RMMI template is proposed to encapsulate the discrete manufacturing data. After the encapsulation, the manufacturing information can be stored under a standard uniform, which can be accessed easily by

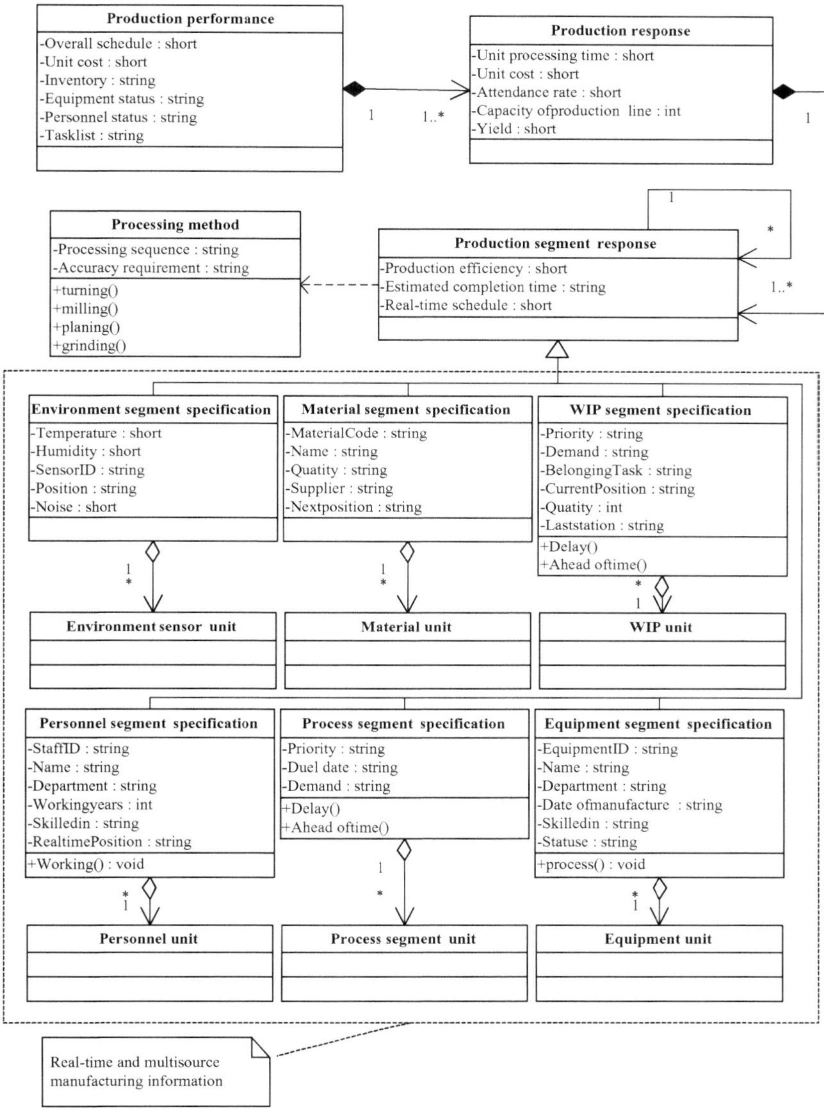

FIGURE 3.4 B2MML-based real-time and multisource manufacturing information template.

the different managers. For that the manufacturing information includes information related to equipment, material, personnel, equipment, environment, and WIP, each of them are defined as an element in the template. Besides, the processing method is added in the template as an element to describe how the objects are processed. The B2MML-based real-time and multisource manufacturing information template is given in Fig. 3.4.

3.6.3 Manufacturing Information Sharing

Manufacturing information sharing module is responsible for sharing the RMMI with third-party applications. As seen in Fig. 3.5, two communication methods are used to share the manufacturing data, namely "Push model" and "Get model." Both of the two models ask the users to register in this system first. Then, the real-time and multisource manufacturing data can be published to the users by the different information communication methods. For Push model, the user needs to submit some basic information of himself, including name, staff ID, telephone number, department, position, the kind of subscribed information, etc. When RMMISS captures the subscribed information, it will send a message which contains standardized real-time information to the relevant staffs through Wi-Fi, GSM, and other communication network. In the Get model, system will send captured information to users at the predefined time points *T*.

3.7 CASE STUDY

3.7.1 Hardware Device

Based on the aforementioned architecture of RMMISS, the hardware device and its main components are designed as shown in Fig. 3.6. Fig. 3.6A presents the concept design, it mainly consists of monitor, industrial computer, heterogeneous interface hub, wireless module, GSM, and multiple sensors. As the name suggests, the monitor is used to display the RMMI. The industrial computer is responsible for connecting and communicating with all kinds of smart objects via wired or wireless methods. The heterogeneous interface hub is used manage

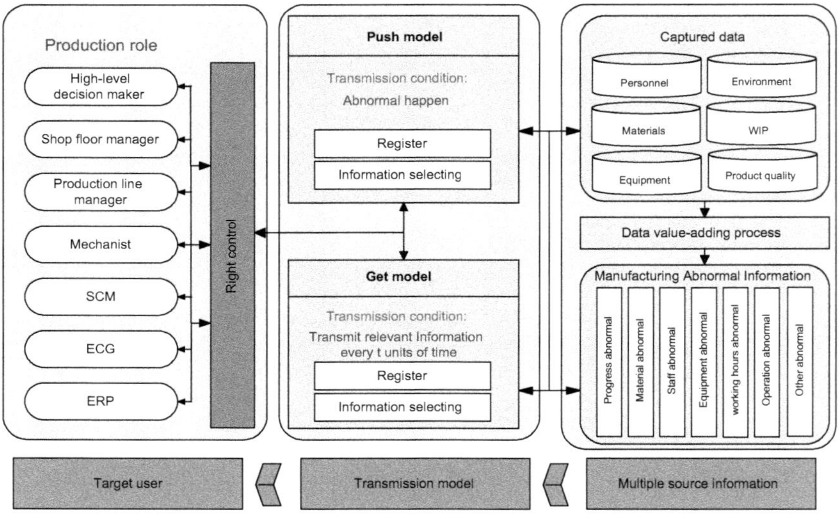

FIGURE 3.5 Manufacturing information transmission mechanism.

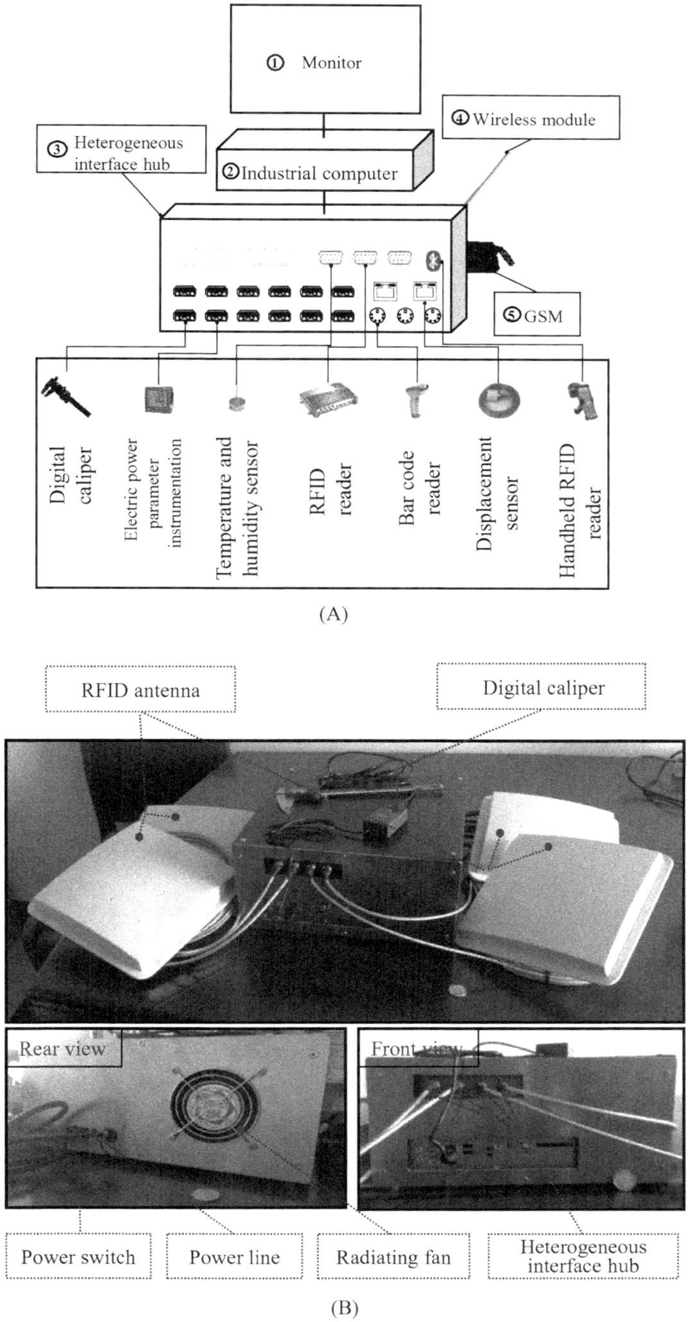

FIGURE 3.6 Hardware device for proposed RMMISS. (A) Concept design. (B) Physical realization.

the multiple sensors through wired way. The wireless module and GSM are used to connect the multiple sensors in a wireless way. For definiteness and without loss of generality, we only choose two types of sensors, namely, RFID and digital caliper, in the physical realization, as shown in Fig. 3.6B.

3.7.2 Software System

In terms of software system, the industrial computer acts as an integrated platform where the relevant drivers of sensing devices and software applications are installed to make the connected sensor devices function normally. To manage the smart sensors in a PnP fashion, the essential procedures are to register the sensors. Thus, a prototype system is designed to establish a bridge for registering and managing the smart sensors. When a new sensor connects to RMMISS, the system are called to register and manage the sensors, the main steps are described as follows:

Step 1: Register a sensor
When a sensor is connected to the RMMISS system through heterogeneous interfaces hub, the system provides user interface to configure it, the equipment name, address, numbers, types, and other necessary attributes are set through the interface, as seen in Fig. 3.7. Besides, if the sensor information needs to be changed, the users can reset the attribute through the information management page.

Step 2: Manufacturing information transmission mode set
As mentioned in Section 3.6.3, the processed RMMI will be sent to the EISs or users through two ways, namely "Push model" and "Get model." This step is used to set the attributes for the two modes, Fig. 3.8 presents the page for information transmission mode set. For the Push mode, the user needs to submit some basic attributes of himself, including name, staff ID, telephone number, department, position, the kind of subscribed information, etc. For the Get mode, the operator needs to provide the information to upload address and time interval, so that the industrial computer can transfer the processed information to the users actively.

Step 3: Manufacturing information display for local users
This step is designed for the local user such as an operator at a single equipment. As some captured data can be used directly by the operator, RMMISS provides visualized way to show the RMMI so that the local operator could find some anomalies as soon as possible, as shown in Fig. 3.9.

Step 4: Manufacturing information storage
The captured raw RMMI will be processed to the meaningful information through the way described in Section 3.6.1 and 3.6.2. In order to achieve the target for tracking and tracing the historical status of manufacturing resources, the information needs to be stored in the database. Fig. 3.10 gives an XML-based manufacturing information storage method.

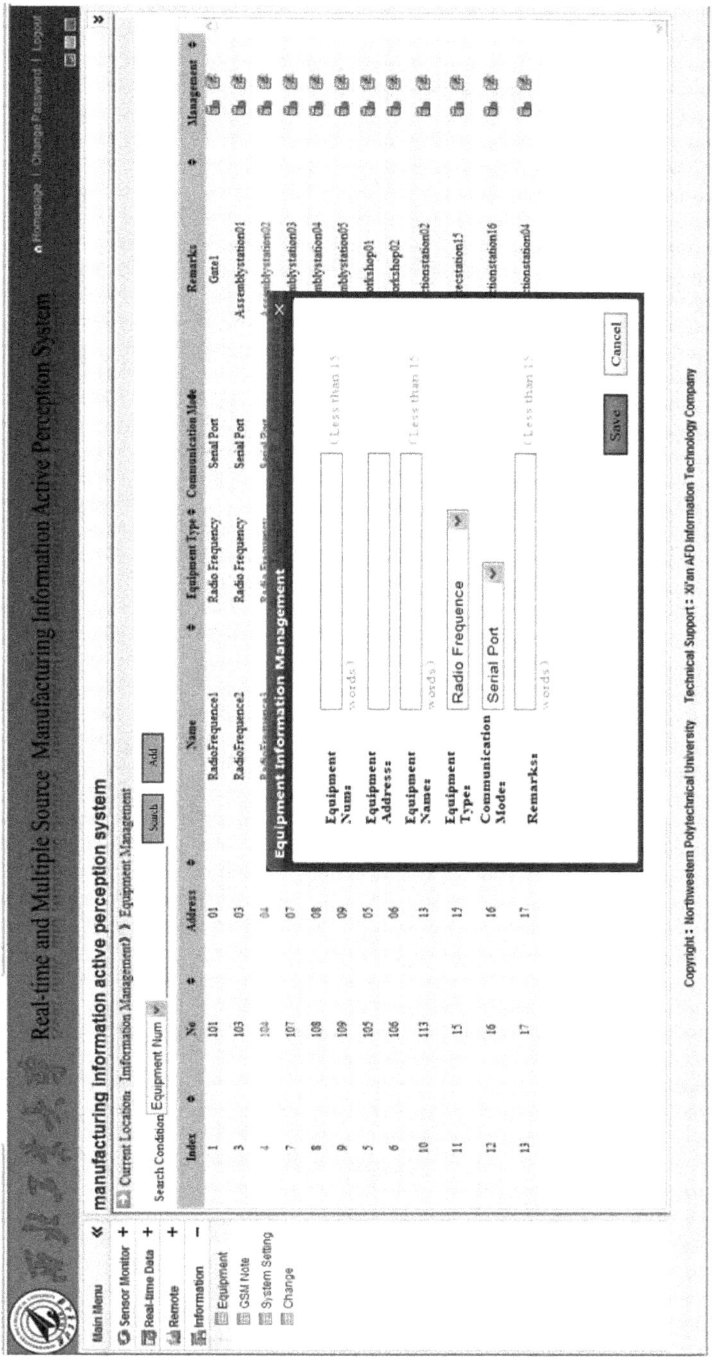

FIGURE 3.7 Sensor registration.

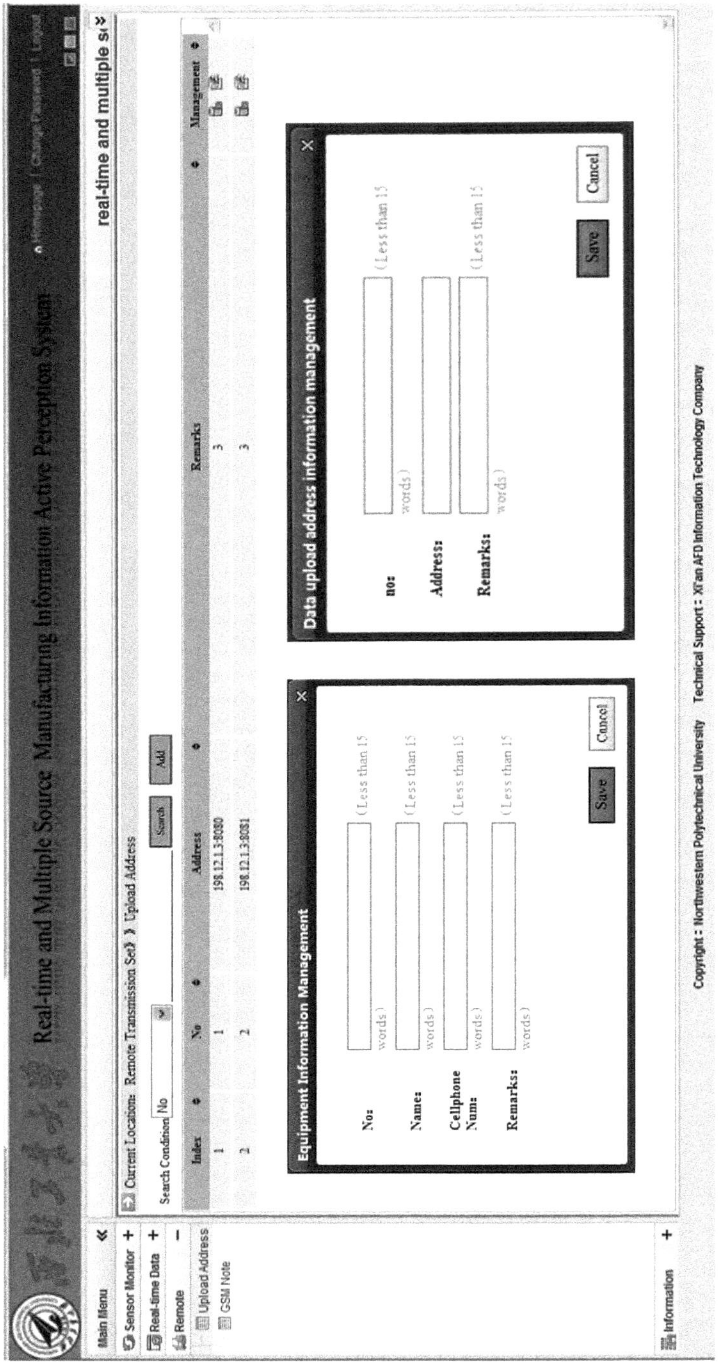

FIGURE 3.8 Manufacturing information transmission mode set.

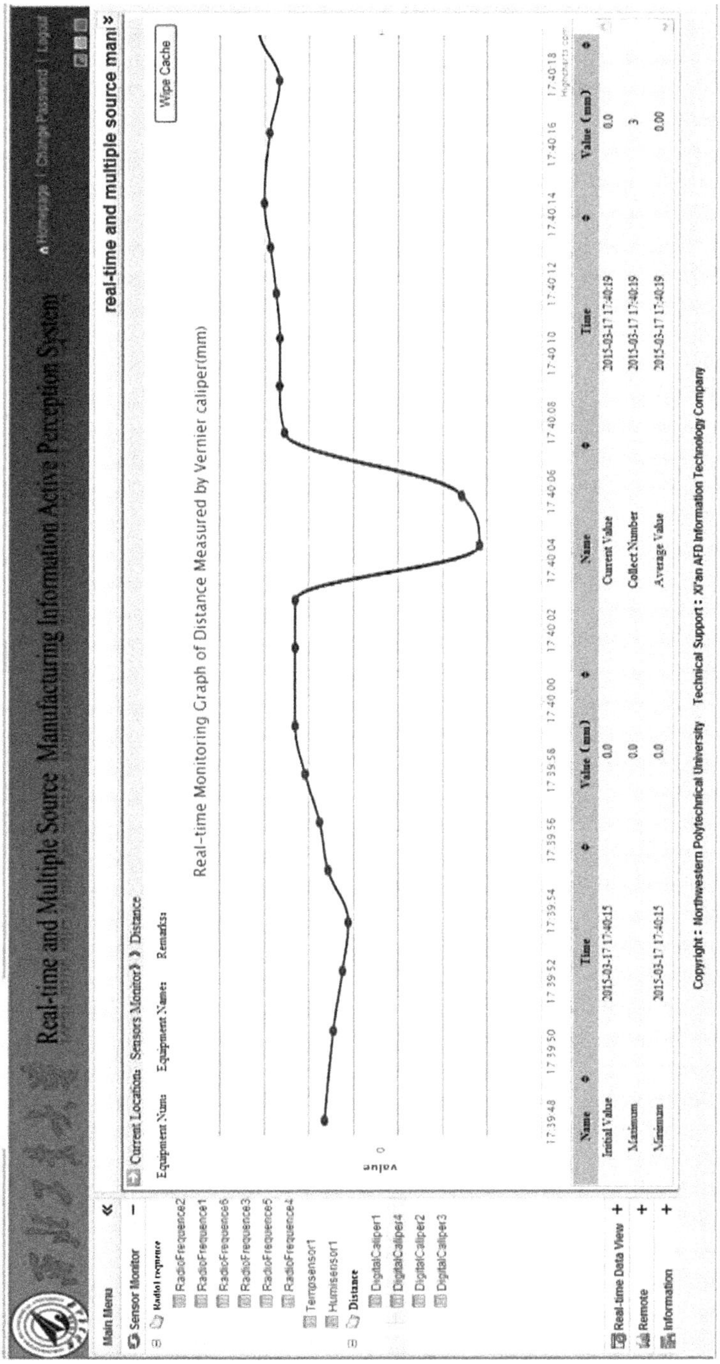

FIGURE 3.9 Manufacturing information display for local users.

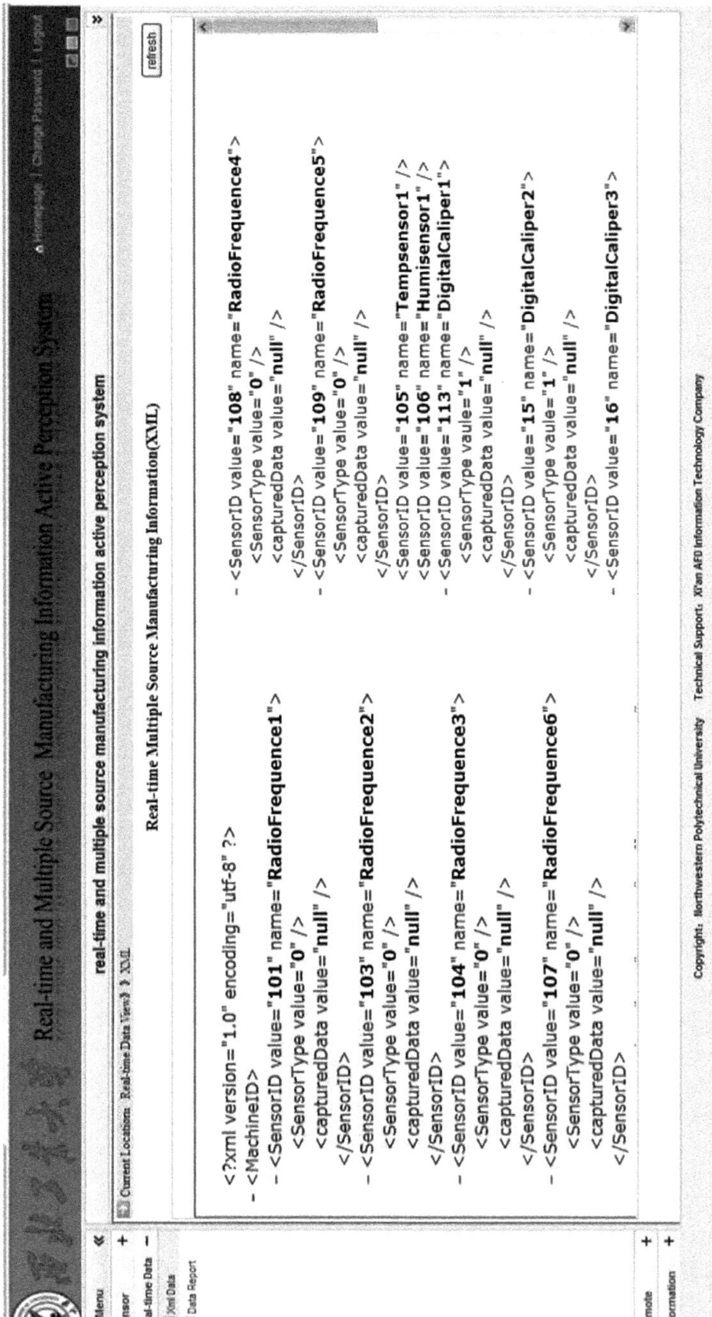

FIGURE 3.10 Manufacturing information storage.

REFERENCES

[1] L. Da Xu, W. He, S. Li, Internet of things in industries: a survey, IEEE Transac. Ind. Inform. 10 (4) (2014) 2233–2243.

[2] Y. Zhang, G. Zhang, J. Wang, S. Sun, S. Si, T. Yang, Real-time information capturing and integration framework of the Internet of manufacturing things, Int. J. Comput. Integr. Manuf. 3052 (December) (2014) 1–12.

[3] S. Zhang, S. Liu, Y. Liu, R. Li, Smart box-enabled product-service system for cloud logistics, Int. J. Prod. Res. (2016) 1–14.

[4] Z. Bi, L. Da Xu, C. Wang, Internet of things for enterprise systems of modern manufacturing, IEEE Trans. Ind. Inform. 10 (2) (2014) 1537–1546.

[5] F. Tao, Y. Zuo, L. Da Xu, L. Zhang, IoT-based intelligent perception and access of manufacturing resource toward cloud manufacturing, IEEE Trans. Ind. Inform. 10 (2) (2014) 1547–1557.

[6] Y. Zhang, J. Xu, S. Sun, T. Yang, Real-time information driven intelligent navigation method of assembly station in unpaced lines, Comput. Ind. Eng. 84 (2015) 91–100.

[7] S.L. Chen, Y.Y. Chen, C. Hsu, A new approach to integrate Internet-of-things and software-as-a-service model for logistic systems: a case study, Sensors 14 (4) (2014) 6144–6164.

[8] W. He, L. Xu, A state-of-the-art survey of cloud manufacturing, Int. J. Comput. Integr. Manuf. 28 (3) (2014) 239–250.

[9] J. Lee, B. Bagheri, H.A. Kao, A Cyber-Physical Systems architecture for Industry 4.0-based manufacturing systems, Manuf. Lett. 3 (2015) 18–23.

[10] D. Ivanov, A. Dolgui, B. Sokolov, F. Werner, M. Ivanova, A dynamic model and an algorithm for short-term supply chain scheduling in the smart factory industry 4.0, Int. J. Prod. Res. 54 (2) (2016) 386–402.

[11] S. Wang, J. Wan, D. Li, C. Zhang, Implementing smart factory of Industrie 4.0: an outlook, Int. J. Distrib. Sens. Netw. 2016 (2016) 1–10.

[12] D. Lucke, C. Constantinescu, E. Westkämper, Smart factory—a step towards the next generation of manufacturing, in: The 41st CIRP Conference on Manufacturing Systems, 2008, pp. 115–118.

[13] X. Yao, H. Jin, J. Zhang, Towards a wisdom manufacturing vision, Int. J. Comput. Integr. Manuf. 3052 (March) (2014) 1–22.

[14] G.Q. Huang, Y.F. Zhang, P.Y. Jiang, RFID-based wireless manufacturing for walking-worker assembly islands with fixed-position layouts, Robot. Comput. Integr. Manuf. 23 (4) (2007) 469–477.

[15] L. Zhang, Y. Luo, F. Tao, B.H. Li, L. Ren, X. Zhang, H. Guo, Y. Cheng, A. Hu, Y. Liu, Cloud manufacturing: a new manufacturing paradigm, Enterp. Inf. Syst. 82 (2014) 167–187.

[16] J. Fang, T. Qu, Z. Li, G. Xu, G.Q. Huang, Agent-based gateway operating system for RFID-enabled ubiquitous manufacturing enterprise, Robot. Comput. Integr. Manuf. 29 (4) (2013) 222–231.

[17] R.Y. Zhong, Q.Y. Dai, T. Qu, G.J. Hu, G.Q. Huang, RFID-enabled real-time manufacturing execution system for mass-customization production, Robot. Comput. Integr. Manuf. 29 (2) (2013) 283–292.

[18] Z. Yang, P. Zhang, L. Chen, RFID-enabled indoor positioning method for a real-time manufacturing execution system using OS-ELM, Neurocomputing 174 (2016) 121–133.

[19] C.K.H. Lee, K.L. Choy, G.T.S. Ho, K.M.Y. Law, A RFID-based Resource Allocation System for garment manufacturing, Expert Syst. Appl. 40 (2) (2013) 784–799.

[20] W.N. Liu, L.J. Zheng, D.H. Sun, X.Y. Liao, M. Zhao, J.M. Su, Y.X. Liu, RFID-enabled real-time production management system for Loncin motorcycle assembly line, Int. J. Comput. Integr. Manuf. 25 (1) (2012) 86–99.

[21] S. Zhou, W. Ling, Z. Peng, An RFID-based remote monitoring system for enterprise internal production management, Int. J. Adv. Manuf. Technol. 33 (7–8) (2007) 837–844.
[22] Y. Zhang, G.Q. Huang, T. Qu, O. Ho, S. Sun, Agent-based smart objects management system for real-time ubiquitous manufacturing, Robot. Comput. Integr. Manuf. 27 (3) (2011) 538–549.
[23] T. Qu, S.P. Lei, Z.Z. Wang, D.X. Nie, X. Chen, G.Q. Huang, IoT-based real-time production logistics synchronization system under smart cloud manufacturing, Int. J. Adv. Manuf. Technol. 84 (1–4) (2016) 147–164.
[24] P.Y. Jiang, Y.B. Fu, Q.Q. Zhu, M. Zheng, Event-driven graphical representative schema for job-shop-type material flows and data computing using automatic identification of radio frequency identification tags, Proc. Inst. Mech. Eng. 226 (2) (2012) 339.
[25] A.W.W. Yew, S.K. Ong, A.Y.C. Nee, Towards a griddable distributed manufacturing system with augmented reality interfaces, Robot. Comput. Integr. Manuf. 39 (2016) 43–55.
[26] A. Deshpande, R. Pieper, Legacy machine monitoring using power signal analysis, in: Proc. ASME 2011 Int. Manuf. Sci. Eng. Conf., 2011, pp. 1–8.
[27] M. Mori, M. Fujishima, M. Komatsu, B. Zhao, Y. Liu, Development of remote monitoring and maintenance system for machine tools, CIRP Ann. 57 (1) (2008) 433–436.
[28] W. Shen, S.Y.T. Lang, L. Wang, iShopFloor: an Internet-enabled agent-based intelligent shop floor, IEEE Trans. Syst. Man Cybern. C 35 (3) (2005) 371–381.
[29] Q. Chi, H. Yan, C. Zhang, Z. Pang, L. Da Xu, A reconfigurable smart sensor interface for industrial WSN in IoT environment, IEEE Trans. Ind. Inform. 10 (2) (2014) 1417–1425.
[30] S. Mayer, R. Verborgh, M. Kovatsch, F. Mattern, Smart configuration of smart environments, IEEE Trans. Autom. Sci. Eng. PP (99) (2016) 1–9.
[31] P. Boonma, J. Suzuki, BiSNET: a biologically-inspired middleware architecture for self-managing wireless sensor networks, Comput. Netw. 51 (16) (2007) 4599–4616.
[32] I.Y. Jung, K.H. Kim, B.J. Han, C.S. Jeong, Hadoop-based distributed sensor node management system, Int. J. Distrib. Sens. Netw. 2014 (2014) 1–7.
[33] S. Wang, W. Su, X. Zhu, H. Zhang, A Hadoop-based approach for efficient web service management, Int. J. Web Grid Serv. 9 (1) (2013) 18–34.
[34] B. Kulvatunyou, N. Ivezic, Y. Lee, J. Shin, An analysis of OWL-based semantic mediation approaches to enhance manufacturing service capability models, Int. J. Comput. Integr. Manuf. 27 (9) (2014) 803–823.
[35] A.A. Fröhlich, L.F. Wanner, Operating system support for wireless sensor networks, J. Comput. Sci. 4 (4) (2008) 272–281.
[36] J. Hu, F.L. Lewis, O.P. Gan, G.H. Phua, L.L. Aw, Discrete-event shop-floor monitoring system in RFID-enabled manufacturing, IEEE Trans. Ind. Electron. 61 (12) (2014) 7083–7091.
[37] L. Minbo, L. Hua, Research on RFID integration middleware for enterprise information system, J. Softw. 6 (2) (2011) 167–174.
[38] V. Anaya, G. Berio, M. Harzallah, P. Heymans, R. Matuleviius, A.L. Opdahl, H. Panetto, M.J. Verdecho, The Unified Enterprise Modelling Language-Overview and further work, Comput. Ind. 61 (2) (2010) 99–111.
[39] M. Pantazoglou, A. Tsalgatidou, A generic query model for the unified discovery of heterogeneous services, IEEE Trans. Serv. Comput. 6 (2) (2013) 201–213.
[40] A. Al-Fuqaha, A. Khreishah, M. Guizani, A. Rayes, M. Mohammadi, Toward better horizontal integration among IoT services, IEEE Commun. Mag. 53 (9) (2015) 72–79.
[41] Y. Luo, L. Zhang, F. Tao, L. Ren, Y. Liu, Z. Zhang, A modeling and description method of multidimensional information for manufacturing capability in cloud manufacturing system, Int. J. Adv. Manuf. Technol. 69 (5–8) (2013) 961–975.

[42] Y. Zhang, G. Zhang, Y. Liu, D. Hu, Research on services encapsulation and virtualization access model of machine for cloud manufacturing, J. Intell. Manuf. (2015).

[43] Y. Zhang, D. Xi, R. Li, S. Sun, Task-driven manufacturing cloud service proactive discovery and optimal configuration method, Int. J. Adv. Manuf. Technol. 84 (1–4) (2016) 29–45.

[44] D. Russom, A. Ahmed, N. Gonzalez, J. Alvarnas, D. Digiusto, Implementation of a configurable laboratory information management system for use in cellular process development and manufacturing, Cytotherapy 14 (1) (2012) 114–121.

[45] T. Kojima, S. Ohtani, T. Ohashi, A manufacturing XML schema definition and its application to a data management system on the shop floor, Robot. Comput. Integr. Manuf. 24 (4) (2008) 545–552.

[46] L.Y. Pang, R.Y. Zhong, J. Fang, G.Q. Huang, Data-source interoperability service for heterogeneous information integration in ubiquitous enterprises, Adv. Eng. Inform. 29 (3) (2015) 549–561.

[47] N. Chungoora, G.A. Gunendran, R.I.M. Young, Z. Usman, N.A. Anjum, C. Palmer, J.A. Harding, K. Case, A.-F. Cutting-Decelle, Extending product lifecycle management for manufacturing knowledge sharing, Proc. Inst. Mech. Eng. B 226 (A12) (2012) 2047–2063.

[48] N. Wu, Z. Mengchu, Intelligent token petri nets for modelling and control of reconfigurable automated manufacturing systems with dynamcial changes, Trans. Inst. Meas. Control 33 (1) (2011) 9–29.

[49] V. Dhokia, X. Zhang, A. Nassehi, S.T. Newman, J. Negre, D. Plakhotnik, B. Lauwers, Development of a data model and a prototype information sharing platform for DEMAT machine tools, Int. J. Comput. Integr. Manuf. May (2014) 1–15.

[50] X. Qiu, H. Luo, G. Xu, R. Zhong, G.Q. Huang, Physical assets and service sharing for IoT-enabled Supply Hub in Industrial Park (SHIP), Int. J. Prod. Econ. 159 (2015) 4–15.

Chapter 4

IoT-Enabled Smart Assembly Station

4.1 INTRODUCTION

Nowadays, manufacturing systems are facing precedent challenges imposed by the highly demanding constraints ranging from the real-time sensing of manufacturing resources to the real-time information sharing among assembly stations, as well as the timely response to the production exceptions and business changes. The current manufacturing systems, especially assembly stations system, based on traditional manufacturing patterns, can hardly respond to the unpredictable exceptions or disturbances in real time, so a new paradigm is urgently needed to meet these challenges better. Currently, the typical challenges that assembly station systems are facing now are: (1) lack of timely, accurate, and consistent information of distributed manufacturing resources of a manufacturing system such as shop floor; (2) lack of an overall solution to track, trace, share the real-time manufacturing information among manufacturing system layer, workshop-floor layer and machine layer, and then provide an optimal decision for the manufacturing system. Recent developments in wireless sensors, communication and information network technologies have created a new era of the Internet of things (IoT). In the recent years, rapid development of advanced technologies such as IoT and the widespread application of these technologies in assembly stations and lines are shifting novel paradigms to meet the demand for proactive perception and dynamic optimization navigation of production processes of smart assembly stations [1–3]. As a result, substantial research efforts have been invested in assembly stations, accompanied with the development of advanced technologies. Assembly lines are flow-line manufacturing systems which are important in the industrial production of high quantity standardized commodities and more recently even in low volume production of customized products [4]. The assembly lines can be classified in two types, namely paced (synchronous) and unpaced (asynchronous) systems according to the classification of assembly systems [5]. For synchronous assembly systems, the whole assembly system is placed by speed of transfer mechanism and the transfer of all assemblies occurs synchronously at fixed time intervals. In asynchronous assembly systems, the assembly operations can move independently

Optimization of Manufacturing Systems Using the Internet of Things
http://dx.doi.org/10.1016/B978-0-12-809910-0.00004-4

and can be queued in front of a workstation. In this research, we focus on the unpaced asynchronous assembly line.

It is well-known that the IoT technology is a promising technology that has been applied in the manufacturing process for sensing and capturing real-time manufacturing resources data. The acquiring and sharing of real-time information of logistics activities is also essential for the collaboration of logistics resources. Zhang et al. proposed an RFID-enabled real-time manufacturing information tracking infrastructure (RTMITI) to address the real-time manufacturing data capturing and manufacturing information processing methods for the extended enterprise. The configuration of IoT environment is the key component of the IoT-based real-time information sensing model [6]. Qu et al. studied the optimal configuration of cluster supply chains with augmented Lagrange Coordination and constructed operation modes and configuration model of cluster supply chains in the alliance leader level, manufacturer level, and supplier level [7]. Zhang et al. describe a framework of the Internet of manufacturing things to share the real-time manufacturing information, and a service encapsulation and virtualization access model to wrap the manufacturing services so that they can be easily reconfigured and reused [8,9]. The relevant researches of RFID technology also describe the model and process of information acquisition, for example, a real-time optimization method for assigning shop-floor material handling tasks and RFID-enabled ubiquitous manufacturing and real-time interactions within production planning and scheduling level [10,11]. Based on the capturing of the dynamic information of relevant resources, instead of directly entering optimization stage, some adaptive controlling mechanisms are entailed. Qu et al. presented a case study of applying RFID for managing material distribution in the complex assembly shop floor of a large air conditioner company [12].

Recently, the rapid progress of information technology (IT) and auto-ID techniques also were applied in strongly supporting to implement the optimal management of real-time manufacturing information. Although the application of these management systems such as lean production (LP), and just in time (JIT) [13,14], etc., brought many creative strategies and models for production management, it was still difficult to solve the existing problems during the production stage. The main reason is that the real-time information of execution in manufacturing process immediately could not be sensed and captured by the upper level management. As a result, the real-time information-driven navigation services for smart assembly stations were hard to achieve. Once exceptions occurred, this would lead to the cascading failures because of the lack of timely, accurate, and consistent manufacturing information and the information-sharing mechanism between the upstream and downstream stations, or even further exacerbating the disturbance of the assembly line. It is essential to design the real-time information-driven navigation services for smart assembly station based on the advanced technologies such as IT, and IoT, and then manufacturers could upgrade the management methods and improve the quality and efficiency of the assembly.

Despite the significant achievements in monitoring and optimization of the assembly process, existing manufacturing paradigms are insufficient to meet requirements imposed by typical challenges and problems in the manufacturing shop floor, especially in assembly station, which includes real-time status monitoring (RSM), real-time production guiding, real-time production data sharing (RPDS), and real-time production requeuing (RPQ). The existing problems for unpaced asynchronous line management are listed as follows.

1. How to monitor the real-time status of smart assembly stations based on the capturing of the real-time information of manufacturing resources, and provide the operators with correct and real-time production guiding to avoid the improper operation and wrong installation of materials.
2. How to actively sense and discover the manufacturing exceptions in the assembly process involving many stations and operators, and achieve the RPDS for timely transmission and handling of manufacturing exceptions and eliminating cascading failures.
3. How to structure a dynamic, collaborative optimization method for the RPQ of tasks' orders based on the real-time status of processes, aimed at achieving efficient control over the whole assembly process and responding the changing production status in real time.

It is challenging for academia and industries to solve these problems. This chapter considers the assembly station as a breakthrough to improve the real-time information-driven control and optimization of assembly process in unpaced asynchronous line. For problems like the RSM, it is important to establish IoT-enabled real-time manufacturing information sensing environment by configuration of advanced automatic identification devices. Then, the RSM for smart assembly stations in the assembly process and real-time production guiding for the upstream and downstream assembly stations could be achieved. Based on the captured real-time information of assembly resources, the RPDS from the bottom layer to the upper layer is designed to transmit and share production manufacturing information. The RPQ based on the Tabu Search (TS) algorithm is used to achieve dynamic optimization of the requeuing of tasks' orders. As a result, the optimal navigation for assembly activities of each station could be achieved. Afterward, the production exceptions and disturb could be easily captured and eliminated among assembly stations and assembly lines, respectively. The proposed architecture and approaches of real-time information-driven navigation for smart assembly station contributes to IoT-enabled process monitoring and control over the assembly lines.

The rest of this chapter is organized as follows. Section 4.2 reviews the related literature. The overall architecture of the IoT-enabled smart assembly station is described in Section 4.3. The RSM is described in Section 4.4. Sections 4.5 and 4.6 illustrate the real-time production guiding and RPDS. Finally, the RPQ is discussed in Section 4.7.

4.2 RELATED WORKS

There are two streams of literature in this research, namely RFID-based applications in assembly line and assistant services for assembly line.

4.2.1 RFID-Based Applications in Assembly Line

The RFID technology has been used mainly in retailing, but there have also been applications to some extent, in the manufacturing environment. Huang et al. report the use of RFID-based wireless manufacturing for walking-worker assembly islands with fixed-position layouts [15]. The emergence of the auto-ID system has helped to mitigate the aforementioned problem by directly indicating to the operators which parts should be installed [16]. However, the performance of the RFID-assisted assembly line is highly dependent on the detection reliability and timeliness of the RFID system. Angeles reported the application of RFID technologies for closing some of the information gaps in the supply chain, especially in retailing and logistics. As a mobile technology, the RFID enables "process freedom" and real-time visibility into supply chains [17]. Gaukler and Hausman performed a thorough analysis of RFID and barcode usage for the avoidance of assembly defects in automotive assembly lines, and they characterized its potential benefits using an analytical model within which several existing barcode implementations were compared [16]. Strassner and Fleisch provided a qualitative overview of the opportunities and potential contributions of RFID technology during assembly in a mixed-model assembly line [18]. Wang et al. presented the deployment of RFID readers and tags to track assembly components on a flexible assembly line. The RFID readers are arranged as a grid at certain positions on the assembly line in order to track any passing components attached with RFID tags [19]. Qiu discussed the need for more effective and efficient factory system integration solutions, suggesting a mechanism to bridge the gap between shop-floor automation and factory information systems utilizing the RFID technology [20]. The RFID captured information is communicated to a cooperating-robots assembly cell for performing welding operations [21–23]. This framework is applied to a typical automotive industry assembly case study. Gwon et al. focused on part delivery and dispatching by integrating RFID technology with a real-time decision support system to ensure the accurate and efficient delivery of auto parts to mixed-model assembly lines [24]. Zhang et al. reported a study on implementing the RFID technology in the application of assembly guidance in an augmented reality (AR) environment. The RFID technology and sensor technology were implemented to detect the operator's activity to facilitate just-in-time information rendering and intuitive information navigation [25]. Scholz-Reiter et al. developed a prototype of a flexible robot-based disassembly cell for obsolete TV-sets and monitors. This system's main parts are a disassembly robot, a handling robot, an intelligent vision system as well as software one for highly flexible online planning and control of the disassembly process. The results lead to optimized reactive planning algorithms as well as to improved sensor systems and the usage of universal and flexible disassembly tools and fixtures [26]. There have been seven possible assembly workflows reported for avoiding assembly defects with RFID, barcodes, or pick-to-light, without considering RFID reading or barcode scanning errors [16]. Huang et al. proposed a wireless manufacturing (WM)-enabled fixed-assembly island with moving operator assembly line. It can be seen from these literatures that dynamic work-in-progress (WIP) information is essential in fixed-position assembly lines because of the existing huge material and personnel movement [27]. Zhang et al. presented a novel research on implementing the RFID technology in the application of assembly guidance in an AR environment. Aiming at providing just-in-time information rendering and intuitive information navigation, methodologies of applying RFID, infrared enhanced computer vision, and inertial sensor was discussed. A prototype system was established, and validated into two case studies. Different from providing guidance for manual operation, Makris et al. presented a RFID-based integration-driven framework for enabling the parts to perform robotic assembly operations in a random mix manufacturing. The RFID infrastructure was applied to sense the newly arriving parts to be assembled and via the integration-driven framework, the robots are able to recognize the parts and perform cooperative operations [28].

4.2.2 Assistant Services for Assembly Line

Assembly lines are the key manufacturing units for many manufacturing systems. The effectiveness and reliability of the assembly has been a researcher question in the industry and academics. The diversified and specialized of consumers' demand for products also led to increasingly complicated and widely ranged assembly process. As a result, in terms of increasing the effectiveness of assembly line, many state-of-the-art literatures were published by scholars in recent years. The current research on real-time assembly guidance system [25,29] mainly relied on the AR systems or virtual reality (VR) systems. These systems have capabilities of providing visual guidance and control over an assembly line for users by the tracking of real-time assembly information, therefore, assembly errors could be reduced or eliminated. Leu et al. also proposed a method for assembly simulation, guidance, planning, and assessment based on the acquired data and the combination of CAD and VR systems [30].

An assembly tree is usually applied to determine the hierarchical assembly sequences in assembly data management. For example, Liverani et al. developed the BAT structure for their assembly sequence check and validation system. Each tree node includes the name and description for the matching component. A rotation and translation matrix is included to indicate its spatial relationship with the subassembly that has been achieved earlier [31]. Yuan et al. developed a visual assembly tree structure (VATS) for their assembly guidance system. Information such as the component name, description, and images can be interactively acquired and rendered to assist the assembly operations. An information authoring interface was also presented. Furthermore, research has been conducted on the interaction between the operator and the system [32].

To have a systemic view of overall performance of the entire system for the engineers and managers, Silva et al. employed computer simulation to integrate with design for manufacturing and assembly (DFMA). However, building a simulation model is tedious, time consuming, error prone, and inefficient [33]. Thus Wy et al. proposed a generic simulation modeling framework to address this research question [34]. An extended formal concept analysis (FCA) was also utilized to analyze relations among patterns of events in historical process logs [35]. In addition, FCA had capability of predicting the probable routes and performances of current progresses periodically. For some complex products that consist of many parts and components or assembly activities, the assembly monitoring and guiding for assembly lines and stations would be prodigious difficulty. For these existing research questions, a subassembly identification method for assembly sequence planning was presented to decompose a complex assembly into a restricted number of subassemblies. As a result, the complexity of assembly could be reduced significantly.

In the traditional manufacturing execution process, the production plans were implemented, the frequent production exceptions (e.g., emergency orders, material shortages, and new requests) can break up the production planning. Therefore, frequent break-up and replanning of production execution resulted in low-efficiency production assembly. To handle these abnormal events efficiently, IoT-enabled real-time production performance analysis and exception diagnosis model based on hierarchical-timed-colored Petri net was established to identify the cause of exception [36]. In a manufacturing environment, face detection techniques are used for two different functions: process fault detection and product quality. Process fault detection is when the measurement systems in an industrial plant are monitored for detecting faulty performance and diagnosing root-causes for process failures. A new fault detection scheme based on the proposed robust one class support vector machine (1-class SVM) is constructed [37]. An evolving approach based on typicality and eccentricity data analytics (TEDA) fault detection in industrial processes was proposed. This approach was a recently introduced algorithm for anomaly detection in data streams [38]. A new use of image processing was presented to detect in real time quality faults and guide manufacturing processes [39]. These proposed approaches provided the feasible solutions for fault detection in a manufacturing environment. A high efficiency assembly planning is a condition precedent in increasing the effectiveness of assembly line. In the recent researches, based on analysis of the requirement of the virtual assembly process planning, a hierarchical assembly task list model was proposed. In the model, assembly tasks were defined to express component assembling operations and were sequentially and hierarchically organized according to different subassemblies, which could perfectly model the construction process of product [40]. A multiplant assembly planning model for generating and evaluating the multiplant assembly sequences was proposed to solve the multiplant assembly planning problem. New graph-based representation models were developed for representing the multiplant assembly sequences [41]. A mathematical programming model was formulated to evaluate all the feasible multiplant assembly sequences, aiming at minimizing the total cost of assembly cost and multiplant cost. Some studies have focused on the assembly line balance problem [42,43]. For example, a mixed-model parallel two-sided assembly line system was proposed to produce large-sized items in an intermixed sequence. A flexible agent–based ant colony optimization algorithm was developed to solve the problem and build flexible balancing solutions suitable for any model sequence launched [44]. The biobjective 0-1 integer programming model was developed to not only solve two-side U-type assembly line balancing problem, but also to solve balancing problems for different design options of two-sided lines [45]. Calleja et al. presented three hybrid metaheuristics, based on simulated annealing, TS and TS with corridor method, to solve the Accessibility Windows Assembly Line Balancing Problem Level2 (AWALBP-L2) for the case where each task can only be performed in one workstation [46]. An effective variable neighborhood search (VNS) was proposed to solve two-sided assembly line balancing problem. In order to achieve a more efficient, balancing, and controllable assembly line, accurate estimate of assembly time was particularly important [47]. There have been various assembly time estimation techniques published by researchers. The multilayer neural networks were presented to approximate the assembly times of two different types of assembly machines based on several parameter combinations. The method was compared with multivariate nonlinear regression for two rather different machine types [48]. Owensby et al. presented a method to estimate assembly times of products. Through structural complexity metric analysis of the graph which was generated from computer aided design (CAD) assembly models, the method used an artificial neural network (ANN) modeler to predict assembly times. The method was shown to be robust and insensitive to different modeling engineers [49].

4.3 OVERALL ARCHITECTURE OF IoT-ENABLED SMART ASSEMBLY STATION

The aim of the research reported in this paper is to provide real-time information-driven navigation services for each assembly station to achieve real-time guidance and optimization. Here, RFID technologies are adopted to collect real-time manufacturing information.

Fig. 4.1 shows the overall architecture of the proposed IoT-enabled smart assembly station, which includes two key sections from the bottom to the top. The bottom section of Fig 4.1 describes that RFID technologies are applied to sense and capture the real-time manufacturing information of operators, assembly processes, and the WIP inventories. Three areas are classified, namely raw

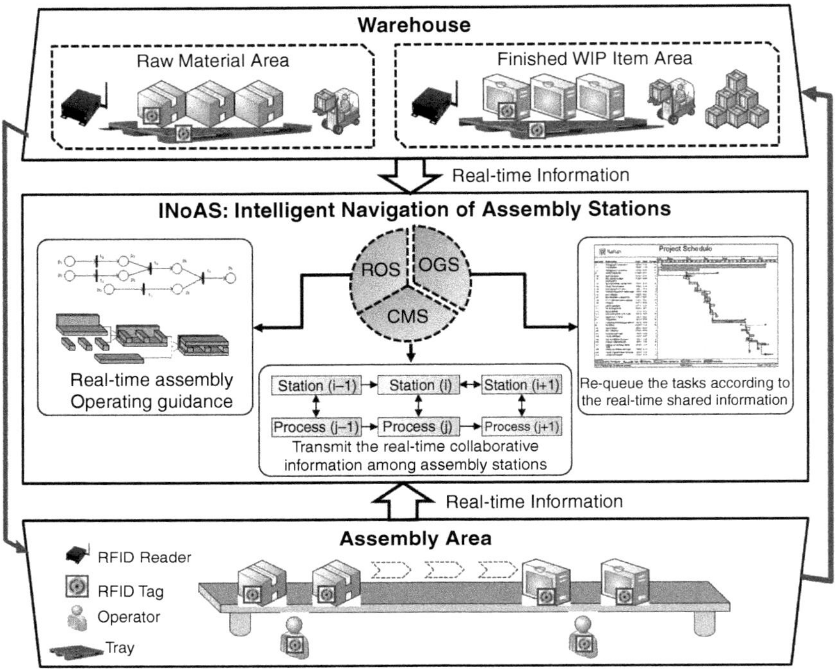

FIGURE 4.1 Architecture of IoT-enabled smart assembly station.

materials area, assembly area, and finished WIP item area. For each area, an antenna is installed to capture the manufacturing items attached with RFID tags. Then, the real-time manufacturing information could be tracked and traced, and it is the key to the designed real-time information-driven navigation services for the assembly stations.

The core services of the IoT-enabled smart assembly station are shown in the top of Fig 4.1. The navigation services are designed. The first service is production guiding (OGS). It is responsible for providing the operation details and instructions during the assembly process to operators, which could reduce the frequency of quality problems as a result of the wrong installation of materials or the improper operation. The second is RPDS. It is used to establish the dynamic connection of manufacturing information between the upstream and downstream assembly stations. The assembly stations can get the collaborative information on other relevant stations. The cooperative information will support the IoT-enabled smart assembly station to rapidly capture the production exceptions and do the right decision making. The third is RPQ. The purpose of it is to queue the order of the tasks of each assembly station based on the captured exceptions such as the lack of raw materials, changed delivery time, new task with high priority, etc. These three kinds of navigation services will be described in the next section in details.

4.4 REAL-TIME STATUS MONITORING

The IoT-based real-time sensing model is the foundation for RSM and intelligent navigation services for smart assembly stations. Configuration sensing devices such RFID devices, PGS locator device, 4G communication device, geographic information system (GIS) are used to build the based-IoT physical sensing architecture for assembly lines and assembly stations. As a result, the status information of assembly operations could be captured and transmitted automatically. The GPS is used to locate the assembly parts' location; RFID information collection devices installed in the assembly lines enable the assembly stations to have the capacity of acquiring its real-time physical status, that is, orders, tasks, volume, list of loading tasks. The 3G communication device is responsible for the both-way transmission of manufacturing information. The GIS is used to offer the optimal path navigation for the driver according to the real-time tasks of the assembly.

Real-time status information of assembly activities is significantly dynamic, so real-time status information updating mechanism is used to renew the assembly tasks, assembly planning, and assembly orders. For example, the real-time values of the assembly progress, the number of assembly parts would be updated once the RFID readers perceive there are tasks moving, the information of the moving task would be input to the updating module. The real-time information sensing and processing system uploads the real-time order information of the assembly sequencing to the GIS. The GIS considers the current assembly condition and applies the path optimization software to offer the optimal path in various forms. Visualization of assembly lines' real-time information is to establish a visualization interface to show the real time information of the assembly. Based on the real-time information of assembly lines and stations, the RSM for smart assembly stations could be achieved.

4.5 REAL-TIME PRODUCTION GUIDING

Fig. 4.2 shows the overall framework of the RPG. The real-time information-driven operating guidance is described in the top of Fig. 4.2. The real-time manufacturing data is captured by the RFID antennas installed at the different areas described in Fig 4.1. The RPG is responsible for eliminating operation errors by providing the operators with correct and visual operation guidance.

To achieve the aforementioned purpose, the Petri net is used as the sequential relationship and constraint of the set of operations related to several factors, such as assembly process planning, and the operator status. The number of materials needed and the process time should also be considered. Therefore, an extension of timed transition color Petri net (TTCPN) is used to represent all the static and dynamical attributes and offer the real-time guidance during the assembly process (Table 4.1).

Definition 4.1: A TTCPN is an 8-tuple $\langle P,T,C,I,O,G,M,D\rangle$

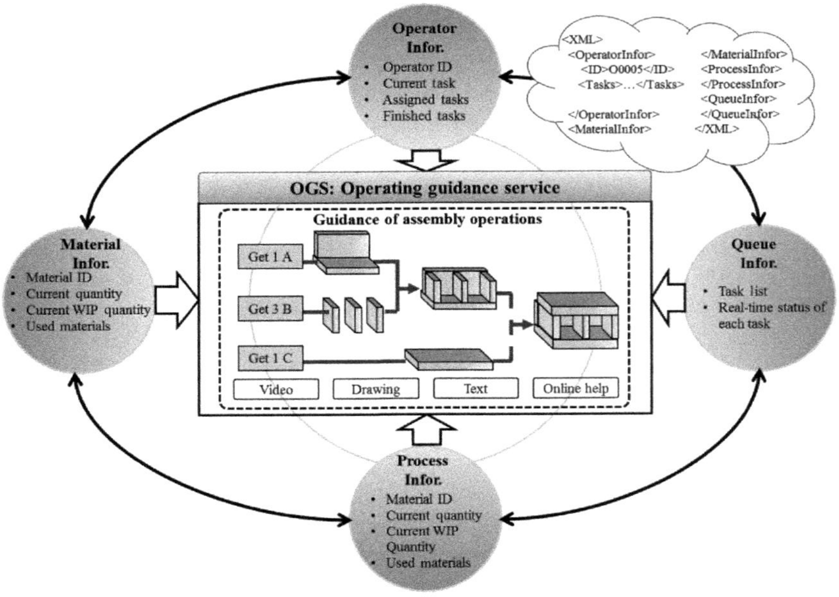

FIGURE 4.2 The real-time production guiding service.

TABLE 4.1 Notations in the Experiments are Defined as Follows

Notation	Description
$P = \{P_1, P_2, \cdots, P_m\}$	A finite set of places
$T = T_1, T_2, \cdots, T_n$	A finite set of transitions
C	Color function from $P \cup T$ to W, where W is some finite set of finite and nonempty sets; an item of $C(S)$ is called a color of s and $C(S)$ is called the color set of S
$I(O)$	Forward (backward) incidence matrix of $P \times T$, where $I(P, T)$ is a function from $C(P) \times C(T)$ to N, where N is the set of nonnegative integers
G	The guard function. It maps each transition t to a Boolean expression (called guard expression)
M	Finite set of tokens which can store time and color information, where M_0 is initial marking of the net which is a vector of P
$D = \{D_1, D_2, \cdots, D_m\}$	The firing time function which assigns the nonnegative (average) firing time, to each transition t of the net

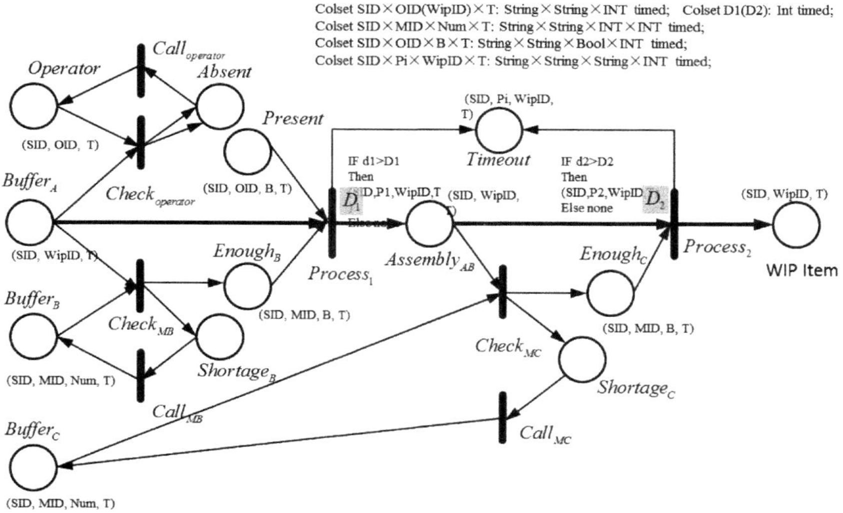

FIGURE 4.3 Petri net–based model of the real-time production guiding.

The TTCPN model is established as in Fig. 4.3 based on the assembly process in the top of Fig. 4.2. The places are drawn as circles, transitions as rectangles or bars, tokens are represented by black dots and arcs by arrows in terms of the graphical representation. The TTCPN model is built up as in Fig. 4.3. According to guard function and consume rule, the position, number, and attributes may vary in the Petri net execution process.

$$\forall P_i \in P : M'(p) = M(p) + O(p,t) - I(p,t) \tag{4.1}$$

When the manufacturing system reaches the time in plan, the begin place acquires a token, and the model starts operating simultaneously. The model checks the operator and the materials in advance before the beginning of each assembly process. Under the IoMT environment, the real-time situation of manufacture resources can be read quickly, the state of tokens will be updated. In order to meet the demand of the navigation activities, the token of operator is in (SID, OID, T) type which contains three parts: [SID = the station ID, OID = the operator ID, T = the read time]. The token of materials is in (SID, MID, Num, T) type which indicates the quantity of the materials. Once the status checks are done, the token of operator will be updated to (SID, OID, B, T) type, the new color variable B is added as BOOL type. (b = "true") shows that the operator is ready, while (b = "false") means unready. The material's tokens will be transferred to the complete place or the incomplete place, if the materials are unready, the tokens will be updated to (SID, MIDs, Short-Nums, T) type, which consists of four parts: SID, MIDs, Short-Nums, and T. The MIDs and Short-nums are composed by three variables to show the shortage conditions.

When ready tokens are acquired, the processes will be executed according to the production processes. Moreover, according to the plan or historical data, the firing time or time delay for processes can be defined, so the exception event can also be acquired.

During the execution, real-time manufacturing information, including the information of operator, material, process, and queue are stored in the dynamical information template as seen in the bottom of Fig. 4.2. This real-time manufacturing information occurring at each station can be easily transmitted and shared to the IoT-enabled smart assembly station of the relative stations through the RPDS for implementing collaborative optimization.

4.6 REAL-TIME PRODUCTION DATA SHARING

The RPDS is used to share the real-time manufacturing information among the upstream and downstream assembly stations. The overall framework of RPDS is shown in Fig. 4.4. Under this framework, each station has the capacity of the real-time inputs and outputs functions through IoT-enabled smart assembly station. The relevant stations can know where inputs are obtained from and outputs are sent to in the execution stage.

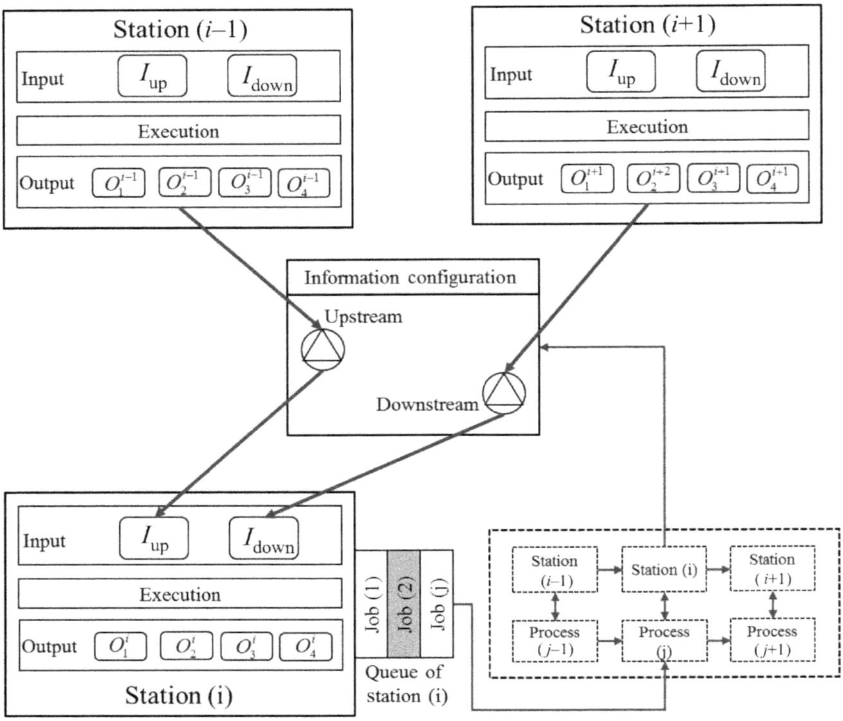

FIGURE 4.4 Implementation framework of the real-time production data sharing (RPDS).

To achieve this purpose, the information of configuration module (ICM) is introduced to define the information flow of the stations. Such relationships can be easily defined in a relational database by ICM. For example, as seen in Fig. 4.4., for each job (*j*) of station (*S*), it is easily to know the relevant upstream station ($s-1$) and downstream station ($s+1$) according to the processing plan and scheduling system. Here, station ($s-1$) indicates that the station is used to execute the previous process ($j-1$) of job (*j*), and station ($s-1$) means the station is used to execute the following process ($j-1$) of job (*j*). For better managing the real-time manufacturing information, for each station, its real-time output is grouped into four classes, namely $\{O_1^i, O_2^i, O_3^i, O_4^i,\}$; and its real-time input is grouped into two classes, namely $\{I_{up}, I_{down}\}$. Here, $O_1^i, O_2^i, O_3^i, O_4^i$ indicate the four kinds of the real-time manufacturing information defined in RPG.

In ICM, for each job (*j*) of the job list at each station (*S*), the real-time information of all the relative stations will be provided as an input to station (*S*). As a result, the input and output relationships of the real-time manufacturing information for all the stations can be built up. Then, the real-time collaborative information of the upstream and downstream assembly stations could be shared. It will provide important input for RPQ.

4.7 REAL-TIME PRODUCTION REQUEUING

The RPQ is responsible for dynamically requeuing the order of the job list at each assembly station based on the shared real-time collaborative information among the relative stations described in RPDS. Compared with traditional dynamical scheduling method of shop floor, the ROS highlights the requeue of each station.

J^s	The process of job (*J*) processed at station (*S*)
p_j^s	Processing time of process J^s
d_j^s	Due date of process J^s
d_j^{s*}	Due date of process J^s before exception occurs
Ts_j^s	Start time of process J^s
Te_j^s	Finished time of process J^s
Td_j^s	Delay time of process J^s
M_s	Station (*S*)
μ_j	Delay penalty per unit time of job *J*
$'d_j^s$	Changes between d_j^s and d_j^{s*}
$Dev = \frac{\sum_j 'd_j^i}{\sum_j p_j}$	Deviation degree of the total changed $'d_j^s$
$\overrightarrow{Dev}$	Limit value of deviation degree *Dev*

In RPQ, each station has *s* jobs in its task queue. Taking station (*S*) as an example, the real-time information of all the upstream and downstream stations related to the s jobs is defined as the input to station (S) in the execution process. As a result, the exceptions (such as the lack of material, the change of delivery time, new insert job, the breakdown of machines, etc.) of any relevant stations can be captured by station (S) in real time. The exceptions of the upstream and downstream stations will lead to a different delayed start time or completion time of the jobs of the queue at station (i). Therefore, it is important to requeue the order of the unprocessed jobs of the job list according to the effect produced by the exceptions of the upstream and downstream stations.

To quickly respond to the exception and get a new optimal job queue, the objective is defined as function (4.2) in this research. It is to minimize the total weighted delay time of all the jobs in the requeued order. Here, the function (4.3) is defined to calculate the delay time (Td_j^s) of each job (j). It means the deviation between the due time (d_j^s) of the job (j) and the finished time of job (j) of the new job queue. The due time of each job (j) is always changed as a result of the exception information of the upstream and downstream stations and is calculated by function (4.4). The functions (4.5) and (4.6) indicate the constraints. Constraint (4.5) describes that the process J^s cannot be processed before its prior process J^{s-1}. "n" is the total number of jobs in the job list at station (S). Constraint (4.6) ensures that only one job can be processed at a time in the same station. In order to simplify the mathematic model, the transport time among two adjacent upstream and downstream processes is neglected.

$$F = \min \sum_{j=1}^{m} \mu_j Td_j^s \tag{4.2}$$

$$Td_j^s = \max(Te_j^s - d_j^s, 0) \tag{4.3}$$

$$d_j^s = \max(Te_j^{s-1} + p_j^s, Ts_j^{s+1}) \tag{4.4}$$

Constraints:

$$Te_j^s - Te_j^{s-1} \geq p_j^s, j = 1, \ldots, n; \forall s \tag{4.5}$$

$$(Te_a^s - Te_b^s \geq p_a^s) \vee (Te_b^s - Te_a^s \geq p_b^s); a, b \in [1, n] \tag{4.6}$$

To address this problem, Tabu Search (TS) algorithm [50], a metaheuristic algorithm, is adopted. The TS algorithm is to search for the next candidate solution from among a carefully constructed neighborhood of the current trial solution, this considers the possibility of the new solution being worse than the existing one [51]. The procedure of the RPQ method is designed in Fig. 4.5 according to the TS algorithm.

In the algorithm, the swap job pair $P(J_x, J_y)$ is generated from current local optimal solution (s_g^i) to its related neighboring solutions through exchanging the related two jobs. As RPQ continually optimizes the queue of the jobs, to

Step1 : Initialization

Generate an initial solution s_k^i, evaluate s_k^i;

Set initial tabu list $tl = \phi$, with its size l;

Set $s_o^i = s_k^i$, $s_l^i = s_k^i$ (s_o^i is the global best solution; s_l^i is the local optimal solution);

Set the maximum number of iteration $N_{\max}$, iteration index $N = 0$;

Step2 : Main Iteration

While ($N \le N_{\max}$)

According to current local optimal solution q_g, generate its related neighbouring solutions, evaluate objective function value of each solution and sort them according to their performances, result as the domain $\{s_1^i, s_2^i, \ldots, s_{m-1}^i\}$, in which s_i^i is better than s_{i+1}^i;

For each s_i^i in $\{s_1^i, s_2^i, \ldots, s_{m-1}^i\}$

If $P(J_x, J_y) \notin tl$

Put $P(J_x, J_y)$ into tl;

If tl is full,

Overflow the earliest $P(J_x, J_y)$ to join the tl;

Update $s_l^i = s_i^i$;

If (s_i^i is better than s_o^i)

Update $s_o^i = s_i^i$;

End For

Else If $P(J_x, J_y) \notin tl$

If s_i^i is better than s_o^i)

Update $s_o^i = s_i^i$, $s_l^i = s_i^i$;

Put $P(J_x, J_y)$ into tl;

End For

End For

$N = N + 1$

End while

Step3 : Output s_o^i.

FIGURE 4.5 Procedure of the requeuing approach in RPQ based on Tabu Search (TS).

improve the solving speed and ensure the quality of the candidate solutions, the first "m" jobs (e.g., $m = 10$) in the job list of station (S) are selected for optimization. Then, the number of the related neighboring solutions generated from each current solution will be "$m-1$." When exceptions occur among the related upstream and downstream assembly stations, the completion time of jobs in upstream stations and the start time of jobs in downstream stations should be changed, which will further lead to the change of due date of the jobs in station (S). The objective station starts to optimize the queue based on the real-time information. It firstly sorts the sequence in ascending order according to the due date of each job and obtains first "m" jobs.

It should be pointed out that the *Dev* in RPQ is used to identify the deviation between the origin scheduler and the captured exception, if the *Dev* can be controlled (e.g., $Dev \leq 15\%$), the RPQ will start. If not, then the exception will be transferred to the upper management system and solved. The RPQ can respond to exception quickly based on the real-time information of the assembly process. The proposed algorithm can rapidly optimize the assembly queue. In general, the RPQ can contribute to reducing or even eliminating the impact of some common exceptions for the stations.

REFERENCES

[1] C. Wang, S. Member, Z. Bi, S. Member, L. Da Xu, S. Member, IoT and cloud computing in automation of assembly modeling systems vol. 10, no. 2, 2014, pp. 1426–1434.
[2] J. Lee, H.-A. Kao, S. Yang, Service innovation and smart analytics for Industry 4.0 and big data environment, Procedia CIRP 16 (2014) 3–8.
[3] T.C. Manufacturing, S. System, F. Tao, Y. Cheng, L. Da Xu, S. Member, L. Zhang, B.H. Li, CCIoT-CMfg: cloud computing and Internet of things, vol. 10, no. 2, 2014, pp. 1435–1442.
[4] N. Boysen, M. Fliedner, A. Scholl, Assembly line balancing: which model to use when?, Int. J. Prod. Econ. 111 (2) (2008) 509–528.
[5] F. Altiparmak, B. Dengiz, A.a. Bulgak, Buffer allocation and performance modeling in asynchronous assembly system operations: an artificial neural network metamodeling approach, Appl. Soft Comput. 7 (3) (2007) 946–956.
[6] Y. Zhang, P. Jiang, G. Huang, T. Qu, G. Zhou, J. Hong, RFID-enabled real-time manufacturing information tracking infrastructure for extended enterprises, J. Intell. Manuf. 23 (6) (2010) 2357–2366.
[7] T. Qu, D.X. Nie, X. Chen, X.D. Chen, Q.Y. Dai, G.Q. Huang, Optimal configuration of cluster supply chains with augmented Lagrange coordination, Comput. Ind. Eng. 84 (2015) 43–55.
[8] Y. Zhang, G. Zhang, J. Wang, S. Sun, S. Si, T. Yang, Real-time information capturing and integration framework of the internet of manufacturing things, Int. J. Comput. Integr. Manuf. 28 (8) (2014) 811–822.
[9] Y. Zhang, G. Zhang, Y. Liu, D. Hu, Research on services encapsulation and virtualization access model of machine for cloud manufacturing, J. Intell. Manuf. (2015).
[10] Y. Zhang, G. Zhang, W. Du, J. Wang, E. Ali, S. Sun, An optimization method for shopfloor material handling based on real-time and multi-source manufacturing data, Int. J. Prod. Econ. 165 (2015) 282–292.
[11] R.Y. Zhong, G.Q. Huang, S. Lan, Q.Y. Dai, T. Zhang, C. Xu, A two-level advanced production planning and scheduling model for RFID-enabled ubiquitous manufacturing, Adv. Eng. Inform. 29 (4) (2015) 799–812.
[12] T. Qu, H.D. Yang, G.Q. Huang, Y.F. Zhang, H. Luo, W. Qin, A case of implementing RFID-based real-time shop-floor material management for household electrical appliance manufacturers, J. Intell. Manuf. 23 (6) (2010) 2343–2356.
[13] D. Levy, Lean production in an international supply chain, Sloan Manage. Rev. 38 (1997) 94–102.
[14] S. Emde, N. Boysen, Optimally locating in-house logistics areas to facilitate JIT-supply of mixed-model assembly lines, Int. J. Prod. Econ. 135 (1) (2012) 393–402.
[15] G.Q. Huang, Y.F. Zhang, P.Y. Jiang, RFID-based wireless manufacturing for walking-worker assembly islands with fixed-position layouts, Robot. Comput. Integr. Manuf. 23 (4) (2007) 469–477.

[16] G.M. Gaukler, W.H. Hausman, RFID in mixed-model automotive assembly operations: process and quality cost savings, IIE Trans. 40 (11) (2008) 1083–1096.
[17] R. Angeles, RFID technologies: supply-chain applications and implementation issues, Inf. Syst. Manage. 22 (1) (2005) 51–65.
[18] E. Fleisch, M. Strassner, The promise of Auto-ID in the automotive industry, Auto-ID Center, Massachusetts Institute of Technology, Cambridge, MA, 2003. Available from: http://www.autoid.org/SC31/clr/200305_3826_Automotive%20Prpsl.pdf#search='RFID%20tags%20in%20the%20automotive%20industry %20in%20PDF/>.
[19] J. Wang, Z. Luo, E.C. Wong, RFID-enabled tracking in flexible assembly line, Int. J. Adv. Manuf. Technol. 46 (1–4) (2009) 351–360.
[20] R.G. Qiu, RFID-enabled automation in support of factory integration, vol. 23, 2007, pp. 677–683.
[21] P.G. Ranky, Collaborative, synchronous robots serving machines and cells, Ind. Robot 30 (3) (2003) 213–217.
[22] R. Buitenhek, Production capacity of flexible manufacturing systems with fixed production ratios, no. 1985, 2002, pp. 203–225.
[23] R. Koeppe, D. Engelhardt, A. Hagenauer, P. Heiligensetzer, B. Kneifel, Robot–robot and human–robot cooperation in commercial robotics applications robot–robot cooperation in industrial production, 2005, pp. 202–216.
[24] S. Gwon, S.-C. Oh, N. Huang, S. Hong, Advanced RFID application for a mixed-product assembly line, Int. J. Adv. Manuf. Technol. 56 (1–4) (2011) 377–386.
[25] J. Zhang, S.K. Ong, a.Y.C. Nee, RFID-assisted assembly guidance system in an augmented reality environment, Int. J. Prod. Res. 49 (13) (2011) 3919–3938.
[26] H. Scharke, A. Hucht, Flexible robot-based disassembly cell for obsolete TV-sets and monitors, vol. 15, 1999, pp. 247–255.
[27] G.Q. Huang, Y.F. Zhang, X. Chen, S.T. Newman, RFID-enabled real-time wireless manufacturing for adaptive assembly planning and control, J. Intell. Manuf. 19 (6) (2008) 701–713.
[28] S. Makris, G. Michalos, G. Chryssolouris, RFID driven robotic assembly for random mix manufacturing, Robot. Comput. Integr. Manuf. 28 (3) (2012) 359–365.
[29] A. Gupta, D. Fox, B. Curless, M. Cohen, DuploTrack: a real-time system for authoring and guiding Duplo Block assembly, 2012, pp. 389–401.
[30] M.C. Leu, H.a. ElMaraghy, A.Y.C. Nee, S.K. Ong, M. Lanzetta, M. Putz, W. Zhu, A. Bernard, CAD model based virtual assembly simulation, planning and training, CIRP Ann. 62 (2) (2013) 799–822.
[31] A. Liverani, G. Amati, G. Caligiana, A CAD-augmented reality integrated environment for assembly sequence check and interactive validation, Concurr. Eng. 12 (1) (2004) 67–77.
[32] M.L. Yuan, S.K. Ong, a.Y.C. Nee, Augmented reality for assembly guidance using a virtual interactive tool, Int. J. Prod. Res. 46 (7) (2008) 1745–1767.
[33] C.E.S. da Silva, E.G. Salgado, C.H.P. Mello, E. da Silva Oliveira, F. Leal, Integration of computer simulation in design for manufacturing and assembly, Int. J. Prod. Res. 52 (10) (2013) 2851–2866.
[34] J. Wy, S. Jeong, B.-I. Kim, J. Park, J. Shin, H. Yoon, S. Lee, A data-driven generic simulation model for logistics-embedded assembly manufacturing lines, Comput. Ind. Eng. 60 (1) (2011) 138–147.
[35] B. Kang, J. Jung, N. Wook Cho, S. Kang, Real-time business process monitoring using formal concept analysis, Ind. Manage. Data Syst. 111 (5) (2011) 652–674.
[36] Y. Zhang, W. Wang, N. Wu, S. Member, C. Qian, IoT-enabled real-time production performance analysis and exception diagnosis model, vol. 13, no. 3, 2016, pp. 1318–1332.
[37] S. Yin, X. Zhu, C. Jing, Fault detection based on a robust one class support vector machine, Neurocomputing 145 (2014) 263–268.

[38] C.G. Bezerra, B.S.J. Costa, L.A. Guedes, P.P. Angelov, An evolving approach to unsupervised and real-time fault detection in industrial processes, Expert Syst. Appl. 63 (2016) 134–144.
[39] F.M. Megahed, J.a. Camelio, Real-time fault detection in manufacturing environments using face recognition techniques, J. Intell. Manuf. 23 (3) (2010) 393–408.
[40] Y. Jun, J. Liu, R. Ning, Y. Zhang, Assembly process modeling for virtual assembly process planning, Int. J. Comput. Integr. Manuf. 18 (6) (2005) 442–451.
[41] Y.-J. Tseng, J.-F. Jhang, F.-Y. Huang, Multi-plant assembly planning models for a collaborative manufacturing environment, Int. J. Prod. Res. 45 (15) (2007) 3333–3349.
[42] G. Tuncel, S. Topaloglu, Assembly line balancing with positional constraints, task assignment restrictions and station paralleling: a case in an electronics company, Comput. Ind. Eng. 64 (2) (2013) 602–609.
[43] J. Sternatz, Enhanced multi-Hoffmann heuristic for efficiently solving real-world assembly line balancing problems in automotive industry, Eur. J. Oper. Res. 235 (3) (2014) 740–754.
[44] I. Kucukkoc, D.Z. Zhang, Mixed-model parallel two-sided assembly line balancing problem: a flexible agent-based ant colony optimization approach, Comput. Ind. Eng. 97 (2016) 58–72.
[45] K. Ağpak, M.F. Yegül, H. Gökçen, Two-sided U-type assembly line balancing problem, Int. J. Prod. Res. 50 (18) (2012) 5035–5047.
[46] G. Calleja, A. Corominas, A. García-Villoria, R. Pastor, Hybrid metaheuristics for the Accessibility Windows Assembly Line Balancing Problem Level 2 (AWALBP-L2), Eur. J. Oper. Res. 250 (3) (2016) 760–772.
[47] D. Lei, X. Guo, Variable neighborhood search for the second type of two-sided assembly line balancing problem, Comput. Oper. Res. 72 (2016) 183–188.
[48] F. Vainio, M. Maier, T. Knuutila, E. Alhoniemi, M. Johnsson, O.S. Nevalainen, Estimating printed circuit board assembly times using neural networks, Int. J. Prod. Res. 48 (8) (2009) 2201–2218.
[49] J.E. Owensby, J.D. Summers, Assembly time estimation: assembly mate based structural complexity metric predictive modeling, J. Comput. Inf. Sci. Eng. 14 (1) (2014) 011004.
[50] F. Glover, Tabu Search—Part II, ORSA J. Comput. 2 (1990) 4–32.
[51] F.F. Choobineh, E. Mohebbi, H. Khoo, Discrete optimization a multi-objective Tabu Search for a single-machine scheduling problem with sequence-dependent setup times, vol. 175, 2006, pp. 318–337.

Chapter 5

Cloud Computing-Based Manufacturing Resources Configuration Method

5.1 INTRODUCTION

Recently, rapid development and growing application of information, sensor, and network technologies are taking place in today's industrial field. For example, the introduction of Internet of Things (IoT) [1], radio frequency identification (RFID) [2] and cloud computing (CC) [3] in the traditional manufacturing has been significantly changing the manufacturing paradigm as well as the business mode. Currently, modern industries are undergoing the transformation from production-oriented manufacturing to service-oriented manufacturing [4]. Providing innovative, high-quality, and sustainable service is crucial for companies, particularly for those small and medium-sized enterprises (SMEs) to survive or remain competitive in the increasing globalization. At the same time, collaboration among companies is another prevalent trend. As the sharing of resources, knowledge, and technologies can lower barriers for companies who suffer the technological limitations and make them more cost-effective in complex manufacturing projects. Finally, a win-win situation is achieved. CMfg, emerging as a new computing and service-oriented manufacturing mode, is promising to reshape the service-oriented, highly collaborative, knowledge-intensive, and eco-efficient manufacturing industry [5]. In recent years, CMfg has been extensively studied, including its concept, architecture, characteristics, and core enabling technologies [4,6,7].

Considered as the manufacturing version of CC [8], CMfg borrows the concept of "Everything as a Service" and treats manufacturing resources and capabilities as services. In the CMfg environment, distributed providers encapsulated their manufacturing resources and capabilities into CMSs and published them into the CMfg platform, in which these services can be managed and operated flexibly and efficiently. Therefore, the sharing, circulation, and collaboration of CMSs are achieved. Through the searching interface, demanders can search and

Optimization of Manufacturing Systems Using the Internet of Things
http://dx.doi.org/10.1016/B978-0-12-809910-0.00005-6

invoke qualified CMSs in a pay-as-you-go manner and an on-demand fashion to meet their customized requirements.

As one of the key issues in CMfg, the resources configuration problem has been widely investigated. In the entire configuration process, three phases that we consider vital and should be tightly coupled with each other are included, they are resource modeling, service discovery, and service optimal selection and composition. However, few existing works studied these three parts jointly, and how to orchestrate them considering their inner relations to offer a practical configuration solution for shop floor resources are the few discussed. Facing the characteristics of the CMfg mode, the resources configuration process encounters the following questions:

1. How to build a cloud machine model that can reveal the real-time production status of machines, with the of IoT and RFID technologies applied in the traditional shop floor?
2. How to register and publish manufacturing services into the CMfg platform, so that they can be efficiently managed and easily accessed in a plug-and-play manner?
3. How to realize efficient service discovery and optimal service selection and composition to enhance the flexibility and agility of the resources configuration, to rapidly respond and satisfy the changing demands?

In this work, three critical technologies and measures are employed to tackle aforementioned questions. Deploying the RFID devices in the shop floor can accurately and timely acquire the multisource production data. With the service-oriented technologies such as the ontology and web services, the cloud machine model can be clarified and constructed. GRA is one of the most popular methods for multicriteria decision making (MCDM). The advantages of using GRA are that its results are based on the original data and the calculation is simple [9] which enables to make quick decisions in the configuration process.

The rest of this chapter is organized as follows: Section 5.2 reviews the related research works. Section 5.3 presents the overall architecture of manufacturing resources configuration method. Section 5.4 builds a cloud machine model for manufacturing machines. Section 5.5 designs the framework of MS-UDDI. Section 5.6 proposes a manufacturing service registration and publication method. Section 5.7 illustrates the task-driven manufacturing service configuration method, including a task-driven service proactive discovery mechanism and a GRA-based evaluation method for service selection.

5.2 RELATED WORKS

Related works to this research are reviewed in three categories. They are cloud manufacturing (CMfg), real-time production information perception and capturing, and cloud service selection and composition.

5.2.1 Cloud Manufacturing

As a promising service-oriented manufacturing paradigm, CMfg has attracted attention from many scholars and extensive investigations have been done. Full-scale sharing, free circulation, and optimal collaboration of manufacturing resources and capabilities are key issues in CMfg. Tao et al. investigated the applications of IoT technologies in CMfg to achieve intelligent perception and access of manufacturing resources [10]. Additionally, combing the CC technology, they proposed a CC-based and IoT-based CMfg service system and its architecture [11]. Resource virtualization and encapsulation are critical for CMfg. Morariu et al. introduced the virtualized MES and shop floor architecture in a private cloud to reduce operational costs and improve the flexibility, agility, and maintainability of the manufacturing system [12]. Liu et al. analyzed the features of resources and constructed a multilevel resource virtualization framework [13]. To achieve the formal description of the manufacturing capabilities in CMfg, Luo et al. established a multidimensional information model of manufacturing capabilities [14]. Xu et al. studied the dynamic modeling of manufacturing equipment capability by establishing the mapping relationship between real-time condition data and the ontology model of equipment capability [15]. Based on the IoT and CC technologies, Zhang et al. presented a service encapsulation and virtualization access model for manufacturing machines [16]. After cloud services published into the CMfg platform, how to realize the efficient discovery of services is another hotspot. Guo et al. proposed an agent-based service discovery framework, within which the task agent and service agent are included. And a structural matching method based on both static and dynamic parameters was studied [17]. Li et al. adopted the Ontology Web Language (OWL) to describe the manufacturing services and based on that a similarity algorithm is proposed. Then, a five-step service matching process has shown its advantages [18]. In the CMfg environment, SMEs are typical participants. Huang et al. discussed the manufacturing resource and capability sharing for SMEs and proposed an SME-oriented CMfg service platform, of which the architecture and key technologies are studied in details [19]. Song et al. designed a CMfg service platform for SMEs and researched some common engines like the intelligent matching engine to support a series of manufacturing business processes in CMfg [20]. Wu et al. applied the concept of CMfg in semiconductor manufacturing operations and proposed a semiconductor industry-oriented architecture for CMfg. Finally, a case study showed the created values for the customers and suppliers [21]. Yang et al. studied how federal resources collaboratively complete large complex projects in CMfg mode. And a large equipment complete service (LECS) collaborative logical framework based on CMfg platform and generalized partial global planning (GPGP) is constructed [22]. With the generation of massive, complex data from the RFID-enabled shop floor in CMfg environment, it is more difficult to extract and process data to make them meaningful. Zhong et al. used a RFID-Cuboid model to reconstruct the raw data according to production logic and time series [23].

5.2.2 Real-Time Production Information Perception and Capturing

The IoT is considered as a part of the Internet of the future and will consist of billions of intelligent communicating "things" [24]. In recent years, IoT technologies such as RFID, sensors and so on have been widely applied in the industrial field and significant contributions have been made. In the shop floor environment, physical manufacturing objects are equipped with RFID devices to become "smart," then the real-time production data can be collected [25]. This promotes the real-time traceability, visibility, and interoperability in shop floor planning, execution, and control [26], enhance the implementation of advanced manufacturing strategies and technologies as well [27]. Yang et al. developed an online sequential extreme learning machine (OS-ELM)-based RFID positioning method in the manufacturing execution system, using RFID to efficiently acquire the real-time data of manufacturing objects (MOs) and the OS-ELM to support real-time data processing [28]. With the support of IoT, Zhang et al. presented a real-time information capture and integration architecture of the Internet of manufacturing things (IoMT) [29]. They designed a ubiquitous shop-floor environment powered by wireless devices like RFID. Then, a framework of multiagent-based real-time production scheduling is proposed to achieve the close loop of production planning and control [30]. To realize the collection and synchronization of the real-time data, Luo et al. created a ubiquitous manufacturing (UM) environment in a hybrid shop floor [31]. Zhong et al. developed an advanced production planning and scheduling (APPS) model for the RFID-enabled real-time ubiquitous environment. The model was tested through four dimensions. They found that the release strategy based on real-time information can reduce the total tardiness effectively and the model was immune to disturbances like defects [32]. Guo et al. proposed an RFID-based intelligent decision support system for real-time production monitoring and scheduling in distributed manufacturing environment. This system was verified to enhance the production efficiency, reduce the production waste and the labor cost as well [33]. Arkan et al. presented a RFID-based real-time location system (RTLS) to obtain work-in-process visibility in manufacturing based on the real-time shop-floor data [34]. Zhang et al. designed an optimization method for shop-floor material handling. And through a case study, the effectiveness of the method was analyzed considering the empty-loading ratio and total distance [35]. Zhong et al. proposed a big data approach to mine trajectory knowledge from the RFID-enabled shop-floor logistics data to support decision makings like logistics planning and scheduling [36]. Qu et al. presented an IoT-enabled real-time production logistics (PL) synchronization system to deal with dynamics occurring in the PL processes [37]. Wang described the significance to implement RFID technology into Norwegian manufacturing industry and developed an intelligent and integrated RFID system to improve traceability and visibility in manufacturing process [38].

5.2.3 Cloud Service Selection and Composition

In the CMfg environment, how to optimally select services from massive candidate services; then how to orchestrate the service compositions, are two critical problems. In fact, the selection of service compositions is far more complex than that of single services. Service composition and optimal selection (SCOS) is a typical multiobjective combinatorial optimization problem [39], Tao et al. investigated the problem with multiple objectives and constraints in CMfg and developed a parallel intelligent algorithm to solve it [40]. Huang et al. established the categories of services and respective quality of service (QoS) indexes for SCOS in CMfg and designed a new chaos control optimal algorithm (CCOA) [41]. Liu et al. proposed a "Multicomposition for each task" (MCET) pattern that integrates the incompetent composite services to execute multifunctionality manufacturing tasks. And a hybrid operator–based matrix-coded genetic algorithm (HO-MCGA) is designed [42]. Liu et al. proposed a synergistic elementary service group–based service composition (SESG-SC) to relax the one-to-one mapping between services and subtasks to enhance the overall QoS level and success rate of service composition [43]. Xiang et al. developed a multiobjective optimization algorithm combining the group leader algorithm (GLA) and the idea of Pareto solution for the SCOS problem based on QoS and energy consumption [44]. Xue et al. proposed a QoS model of the service composition considering the horizontal collaboration between services in cluster supply chains (CSC) and designed a genetic-artificial bee colony (G-ABC) algorithm to efficiently find the optimal solution [45]. Zheng et al. presented a design preference-based QoS evaluation method for resource service selection and adopted the particle swarm optimization (PSO) algorithm to select the optimal service composition [46]. The correlations among services may affect the aggregation QoS level of service compositions. Tao et al. took service correlation into consideration for the multiobjective MGrid resource service composition and optimal-selection (MO-MRSCOS) and proposed a PSO-based method for solving this problem [47]. Jin et al. built a correlation-aware service description model to characterize the QoS dependence among services. Based on this, they presented a service correlation model to automatically obtain QoS values of services. Finally, the correlation-aware service optimal selection is solved by the genetic algorithm (GA) [48]. Transportation is another factor influences the QoS property of service composition, which primarily impacts the total service time and cost [49]. Lartigau et al. analyzed the geoperspective correlation among services and developed an adapted Artificial Bee Colony (ABC) algorithm with an initialization enhancement to optimize the computational time [50]. Xiang et al. introduced the "Big manufacturing data" in CMfg, which have brought difficulties and challenges to the SCOS problem. For solving this problem, they presented a case library-based initialization method for the optimization algorithm [51].

5.3 OVERALL ARCHITECTURE OF MANUFACTURING RESOURCES CONFIGURATION METHOD

As shown in Fig. 5.1, the overall architecture of manufacturing resources configuration is illustrated. It consists of four modules which are the servitization of shop-floor resources, manufacturing task publishing, task-driven service proactive discovery, and service optimal configuration, respectively.

The servitization of shop-floor resources, that is, the manufacturing machines, lays the basis of resource sharing and circulation in the CMfg environment. So that demanders can easily search and invoke resources on their need, either in the way as individuals or as the cloud service composition (CSC) when coping with complex tasks. To realize the aforementioned goals, the scientific description of resources needs in-depth research. In resource modeling, several points should be made clear. They are the basic description of the resource, manufacturing capabilities, real-time status information captured by RFID devices and other sensors, and the historical evaluation information. Based on the information model of resource, adopting the ontology and semantic web technologies, the ontology model is built accordingly. Then, through the manufacturing service universal description, discovery and integration (MS-UDDI), CMSs are registered and published into the manufacturing cloud. Meanwhile, with the efficient management of cloud services, newly published tasks in the demand cloud can be rapidly responded.

The task-driven service proactive discovery mechanism turns around the situation that CMSs can only be passively found by demanders. Through this mechanism, service providers are able to make the fast response to the published tasks and request for undertaking corresponding ones proactively according to their real-time production status. By employing the semantic matching method between the functional information of CMSs and task requirements, the qualified CMSs are reserved and then the CMS candidate sets (CMSCSs) are quickly formed. Therefore, the efficiency of resources configuration can be highly enhanced.

The service optimal configuration method is designed to help demanders find satisfactory solutions from the huge numbers of candidates. Considering the real-time status data and evaluation information, we built an evaluation system, in which criteria such as cost, time, QoS, and energy consumption are defined. Additionally, a GRA-based comprehensive evaluation method is proposed to select optimal CMSs and optimize the service compositions as well. Finally, the optimal CMS composition solution is achieved.

5.4 CLOUD MACHINE MODEL

Building a cloud machine model is divided into two steps. First is to build an information model to clearly describe the manufacturing service in four dimensions. Second is to semantically represent the service using ontology and web service technologies, that is, to construct an ontology model of manufacturing service.

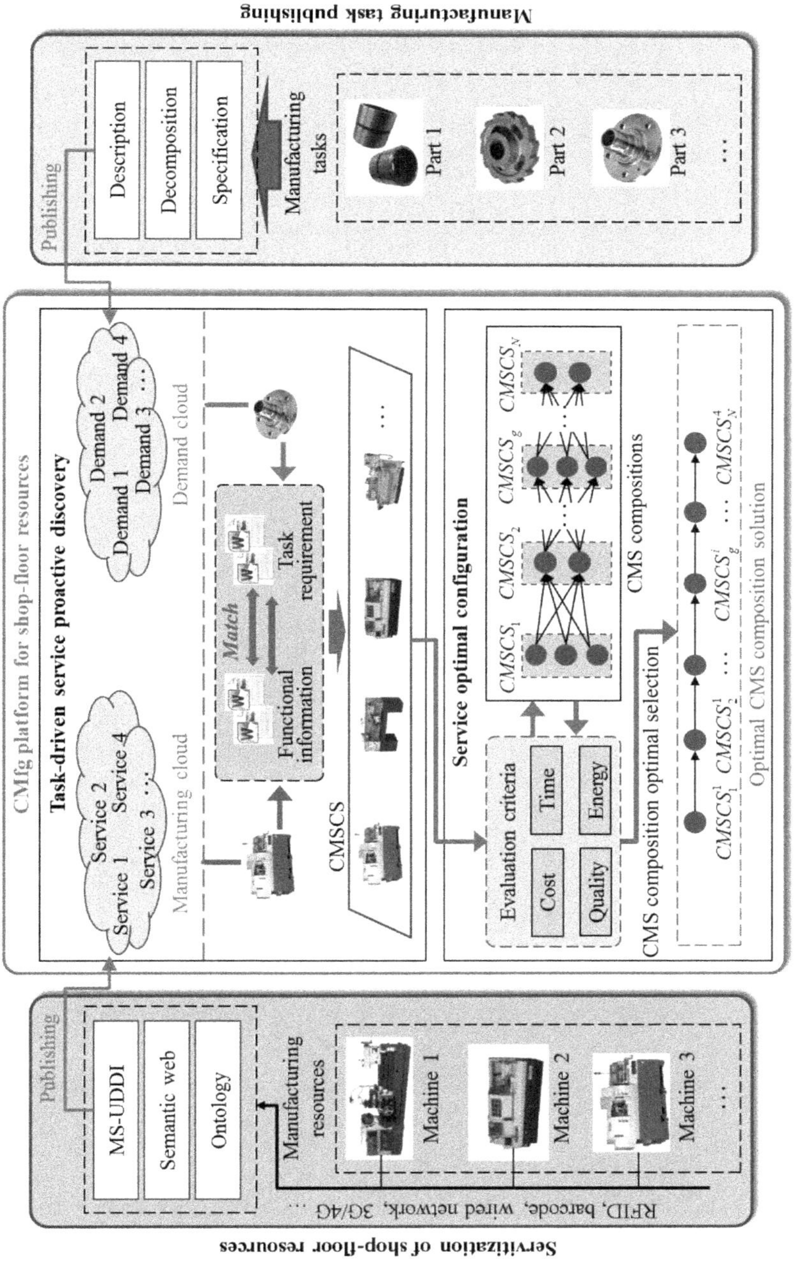

FIGURE 5.1 Overall architecture of manufacturing resources configuration method.

5.4.1 The Information Model of Manufacturing Service

In this section, we studied the information model of cloud machine with both its static description information and dynamic manufacturing information considered. As shown in Fig. 5.2, the cloud machine model comprises four kinds of information, including the basic information, the function information, the real-time status information, and the evaluation information. It is defined as:

$$CMS = (CMBasicInfo, CMFunctionInfo, CMStatusInfo, CMEvaluationInfo).$$

1. Basic information
   The basic information of a CMS describes the inherent attributes of the cloud machine, such as service ID, service name, and so on. Here, service ID is the identification information of a CMS in CMfg platform, which enables the fast positioning of this service when it is searched or invoked by demanders. Some other information includes workshop, manufacturer, purchase date, and service life. It is defined as:

$$CMBasicInfo = (CMID, CMName, CMWorkShop, CMManufacturer, CMPurDate, CMLife).$$

2. Function information
   The function information describes the specific functional attributes of a cloud machine, which supports the executions of service searching and service matching in CMfg. According to the production process, some typical functional attributes are exacted. They are the part type, machining method,

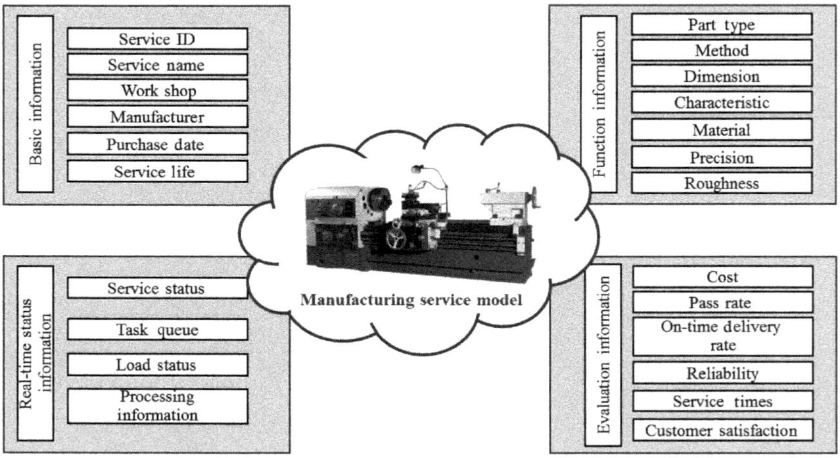

FIGURE 5.2 **Information model of cloud machine.**

geometric dimension, characteristic, machining material, machining precision, and roughness. It is defined as:

$$CMFunctionInfo = (CMPartType, CMMethod, CMDeoDimension, CMCharacteristic, CMMaterial, CM\,Precision, CMRoughness).$$

3. Real-time status information
During the manufacturing process, the real-time status data is accurately acquired by the deployed RFID devices and various sensors like RFID readers, digital calipers. The production data makes the traditional manufacturing activities more transparent, traceable, and controllable, so that it helps upper-level manager be aware of the real-time production status. Furthermore, based on the data, some dynamic optimization methods like real-time production scheduling and online quality control are developed. It is defined as:

$$CMStatusInfo = (CMStatus, CMTaskQueue, CMLoad, CM\,Proces\sin gInfo).$$

4. Evaluation information
The evaluation information consists of two parts; some objective indicators are extracted from the historical production records, such as cost, pass rate, on-time delivery rate (OTDR), reliability, and service times in the CMfg platform; while other subjective ones are rated by customers, here, the customer satisfaction (CS) is considered. When in service evaluation, demanders refer to the evaluation information and choose several criteria that they are most concerned about to find their satisfactory CMSs. It is defined as:

$$CMEvaluationInfo = (Cost, PassRate, OTDR, Reliability, STimes, CS).$$

5.4.2 The Ontology Model of Manufacturing Service

The efficiency and QoS discovery and service matching greatly depend on how to describe the manufacturing services. In this work, ontology and semantic web technologies are employed for achieving aforementioned goals. The semantic web provides a common framework that enables the sharing and reuse of data across the applications, enterprises, and communities [52]. Ontologies can not only explicitly represent the domain knowledge and clarify their relationships but also have strong reasoning ability. To effectively express the manufacturing information as well as the connotative meanings, the ontology description language OWL-S is utilized to describe the ontology model. As OWL-S can semantically describe the web services according to the capabilities offered and perform logic inference to service match between the offered capabilities and the required capabilities [53].

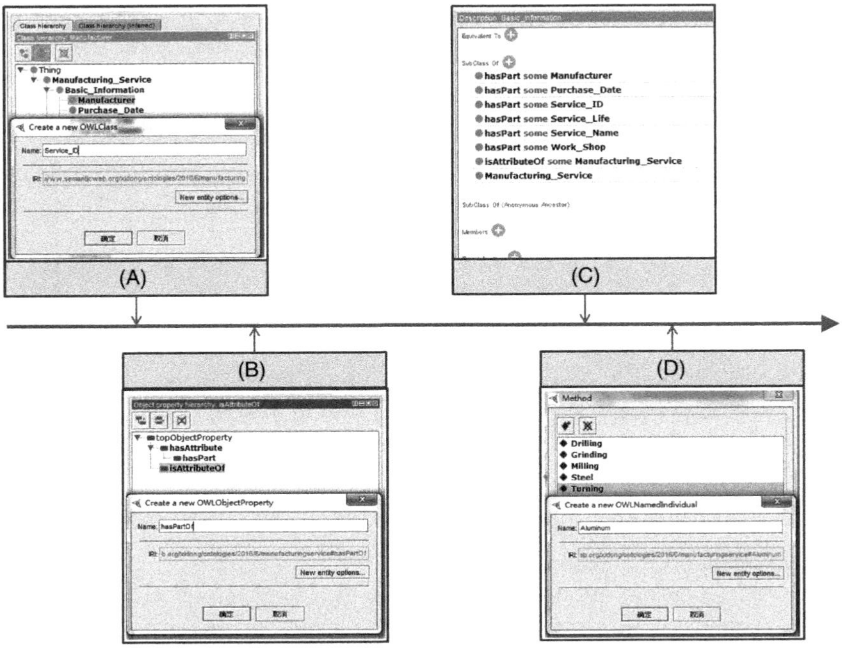

FIGURE 5.3 Steps for constructing an ontology model of manufacturing service. (A) Define classes. (B) Define properties. (C) Describe classes. (D) Create individuals.

Generally, the ontology development process consists of following steps: defining the class and the class hierarchy, defining the property set with object properties and data properties included, and creating individuals. In this work, the Protégé 4.3 is used to construct the ontology model of manufacturing service. As shown in Fig. 5.3, the main process to develop the ontology is illustrated. In the first step, we create the concept classes of manufacturing service and define the class hierarchy which is shown as a structure tree in Fig. 5.3A. The second step is to define the object properties between two individuals in different classes. In Fig. 5.3B, two pairs of inverse properties are defined, namely "hasAttribute" and "isAttributeOf," "hasPart," and "isPartOf." Third, the relations between classes are described by the defined properties. The description of "Basic_Information" is shown in Fig. 5.3C. Finally, Fig. 5.3D shows how to create individuals that belong to the corresponding classes. Here, for example, individuals belonging to "Method" are like drilling, grinding, milling, and turning. "Material" comprises steel, copper, aluminum, and cast iron. Repeat aforementioned steps until all the classes and relations are defined. Fig. 5.4 is a screen shot of the established ontology model of manufacturing service. As depicted, part 1 is the complete structure tree of concept classes; part 2 shows the ontology relation graph of manufacturing service.

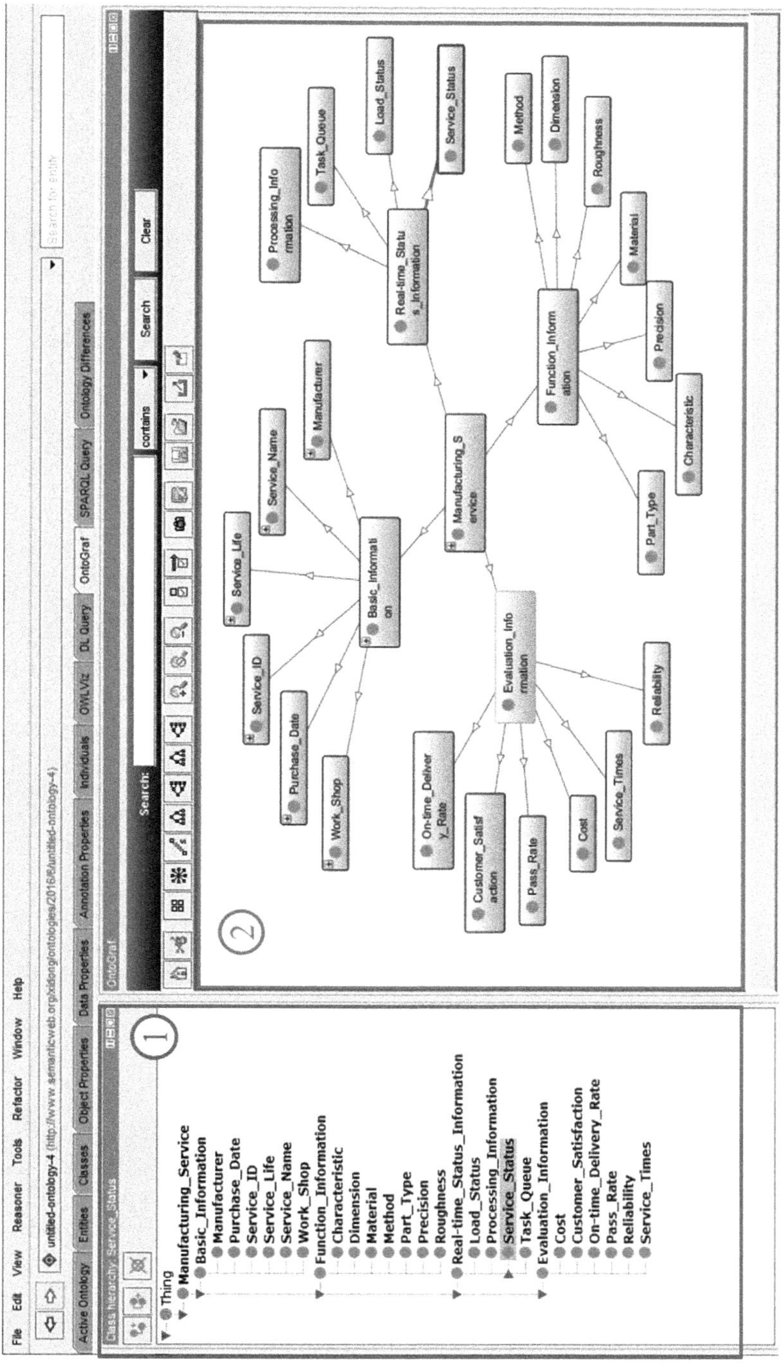

FIGURE 5.4 Ontology for manufacturing service.

5.5 MS-UDDI

In this section, we developed a MS-UDDI-based registration and publication method for manufacturing services accessing to the CMfg platform. What will realize the full sharing of services, and then enhance the utilization of services. First, the UDDI is briefly introduced. Then, the framework of MS-UDDI is illustrated.

5.5.1 UDDI

Universal description, discovery and integration (UDDI) is a specification for web-based information registries of web service. It is also a publicly accessible set of implementations of the specification that define a way for businesses to register and publish the web services they provide, so that other businesses can discover them [54].

In the UDDI registries, the core information model is defined in an XML schema that describes four types of data structures. They are businessEntity, businessService, bindingTemplate, and tModel.

1. businessEntity: It describes the overall information about services that businesses offer; it mainly includes the business name, contact information, categorization, and some other key identifiers. All these information helps other businesses to search and locate the businesses that provide a particular web service.
2. businessService: It is a descriptive container of a series of related services provided by businessEntity, which includes the information about the business processes and taxonomical category of services; and a businessEntity may contain one or more businessServices.
3. bindingTemplate: It is the technical web service description that defines the required information when invoking specific web services; within a businessService, there exist one or more bindingTemplates.
4. tModel: It is a list of key references contained in each bindingTemplate and serves as a pointer to the information about specifications. It is metadata that mainly includes the service name, publishing organization, and the URL pointers to the specifications.

5.5.2 The Framework of MS-UDDI

Based on the UDDI technology, we built a MS-UDDI to realize efficient and integrated management of manufacturing services; to allow service providers to register and publish their manufacturing services into the CMfg platform. Therefore, these services can be easily found and invoked by demanders. As shown in Fig. 5.5, the framework of MS-UDDI consists of three submodules, including the registration module, the publishing module, and the search module.

In the registration module, service providers can register and publish their manufacturing services into the MS-UDDI through the service registration graphical user interface (GUI). The detailed information of services such as

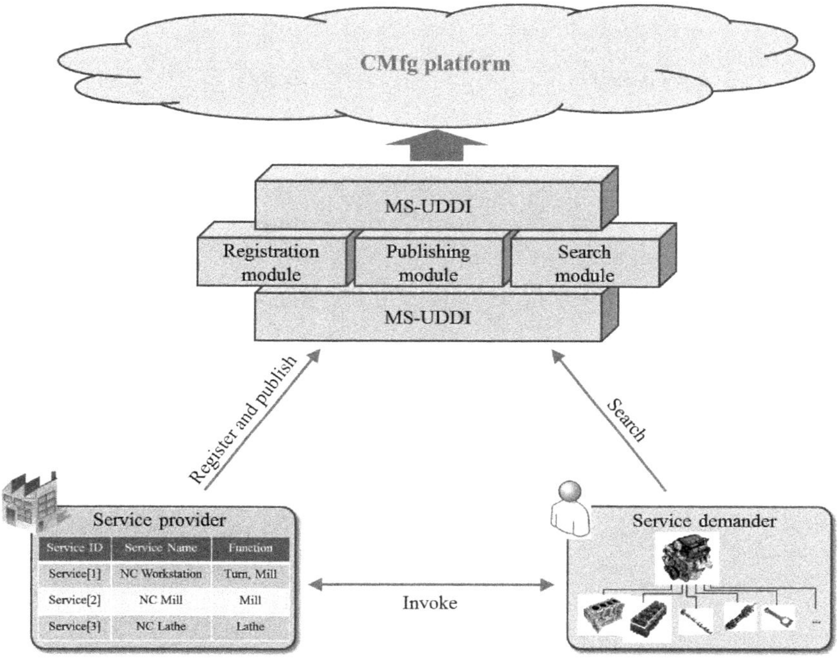

FIGURE 5.5 The framework of MS-UDDI.

name, category, and location are provided in the registration process. And each service is given a universally unique identifier (UUID) that makes the service traceable and trackable. The registered services are accurately classified and aggregated through multiple dimensions like service type, region, and so on. In the publishing module, services are published into the CMfg platform so that large-scale sharing of distributed services can be achieved. Meanwhile, via MS-UDDI, various services are freely circulating in the cloud environment, which makes it convenient for demanders to search and find services that support the related manufacturing activities, even the entire production life cycle. Aforementioned procedures are executed in the search module. After demanders find their target services, the business contracts between two parties are signed. During the service execution, demanders or other users can invoke the binding services to acquire corresponding real-time production information at any time.

5.6 MANUFACTURING SERVICE REGISTRATION AND PUBLICATION

In MS-UDDI, to complete the registration of manufacturing services in the registration module, we have to embed the OWL-S profiles that describe the services into a UDDI data structure. One way to achieve that is through the

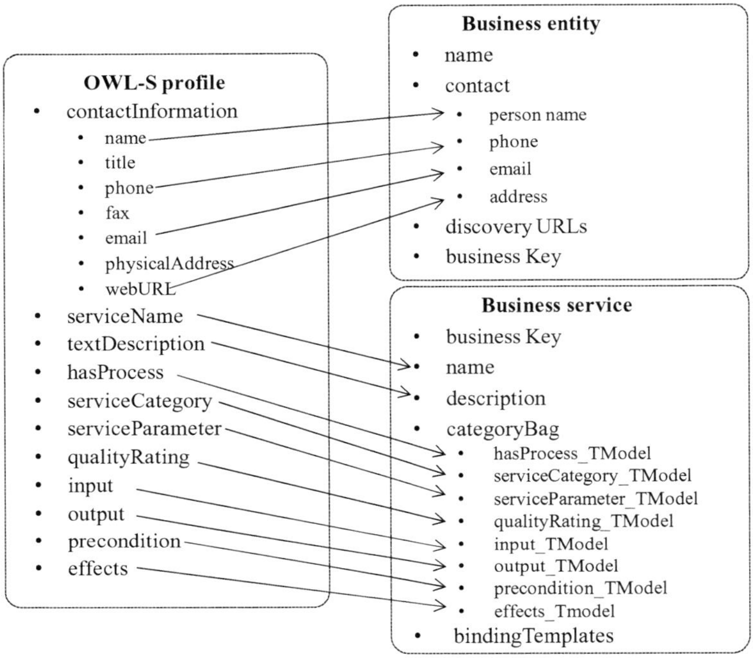

FIGURE 5.6 Mapping between OWL-S profile and UDDI [53].

mapping approach presented by Srinivasan et al. [53]. As shown in Fig. 5.6, if an element in the OWL-S profile can find a corresponding one in UDDI, a one-to-one mapping relationship is built, such as contactInformation and serviceName in the OWL-S profile. If not, a tModel-based mapping method is used. Specialized UDDI tModels are defined for all unmapped elements in the OWL-S profile like serviceCategory and serviceParameter. Finally, all of the OWL-S profile elements are converted into the UDDI elements. In the registration procedure, service providers first register manufacturing services into the MS-UDDI. Then, some other application services that encapsulated based on the real-time production data for supporting the manufacturing processes are registered, such as online quality controlling service and real-time production scheduling. Therefore, when demanders invoke the binding services, they can not only acquire the real-time data but also further analyze the execution status to meet their demand with the application services working as function tools.

In the publishing module, a web server should be installed with related web service components plugged in. Therefore, the registered manufacturing services are deployed on the web server. If successfully deployed, the Web Services Description Language (WSDL) files can be viewed. This service deployment procedure makes services accessible over the Internet. After being published into the CMfg platform, services are then pooled into the corresponding manufacturing clouds and each of them can be regarded as a droplet in the cloud, that is, the CMS.

The search module enables demanders to search and inquire CMSs on their needs. Through the service searching GUI, demanders can either input several keywords to describe their requirements or directly upload the task files in the standard paradigm. The core of the service discovery is a similarity matching algorithm that finds the CMS with the highest similarity to the description of requirements. As a searching result, the UUIDs referencing the competent CMSs are sent to demanders for invocation. Meanwhile, the invocation portal and the required input parameters are also provided in the searching results. Accessing this portal, demanders can remotely acquire the real-time production data and interact with providers as well.

5.7 TASK-DRIVEN MANUFACTURING SERVICE CONFIGURATION MODEL

This section illustrates a task-driven configuration model for manufacturing machines in shop floor when dealing with part-level tasks. Let $MT = \{ST_1, ST_2, ..., ST_g, ..., ST_N\}$ denote the part-level task under processing, where N is the total number of subtasks and ST_g is the gth subtask. The service optimal configuration process for this task follows the next two parts.

5.7.1 Task-Driven Service Proactive Discovery

In today's manufacturing industry, distributed SMEs are insufficient to find tasks proactively, which results in the low utilization of manufacturing resources. To improve this, we presented a task-driven service proactive discovery mechanism that enables CMSs to actively make rapid responses to tasks and then apply to perform corresponding ones. Once applications are made, a semantic-based intelligent match from the functional perspective is performed between CMSs and tasks to select competent services. As shown in Fig. 5.7, the semantic matching is performed considering all the functional attributes of CMSs.

Here, based on the semantic match method [55], four degrees of matching are assigned, including *Exact*, *Plug in*, *Subsume*, and *Fail*. According to Eq. (5.1–5.7), all degrees of matching are measured in the matching procedure.

$$Match(CMPartType, MTPartType) = Exact \tag{5.1}$$

$$Match(CMMethod, MTMethod) = Exact \tag{5.2}$$

$$Match(CMGeoDimension, MTGeoDimension) = Exact \tag{5.3}$$

$$Match(CMCharacteristic, MTCharacteristic) = Exact \tag{5.4}$$

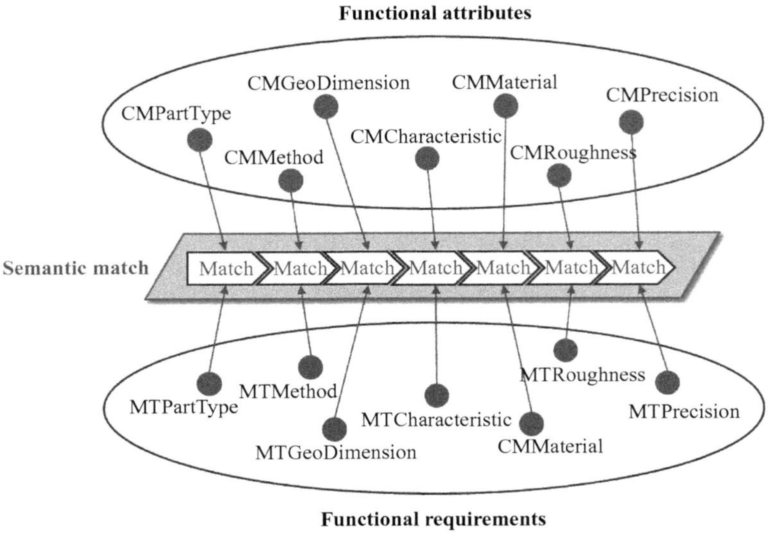

FIGURE 5.7 **Semantic matching method.**

$$Match(CMMaterial, MTMaterial) = Exact \tag{5.5}$$

$$Match(CMPrecision, MTPrecision) = Exact \tag{5.6}$$

$$Match(CMRoughness, MTRoughness) = Exact \tag{5.7}$$

And only if all the degrees of matching reach *Exact*, this CMS can be viewed as competent to undertake the task and pooled into the candidate set. The strategy of service proactive discovery can rapidly discover potential services that satisfy the task requirements, thus greatly reducing the service response time which is one of the key indicators for measuring the efficiency of resources configuration.

5.7.2 Service Optimal Configuration Method

CMSs in the candidate set offer the same kind of functional services to a specific demander but still differ in many other service characteristics like cost, time, and so on. How to choose a CMS to best satisfy the demander's customized requirements is studied in this section. A service optimal configuration method is proposed aiming to find an optimal solution for the task. The specific procedure is shown in Fig. 5.8, in which an evaluation system and a GRA-based evaluation approach are included.

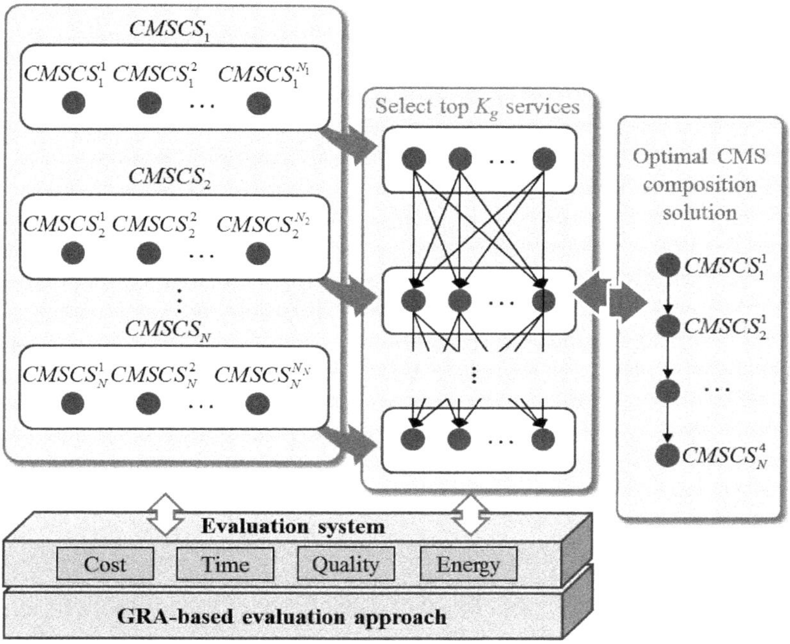

FIGURE 5.8 Service optimal configuration method.

In the evaluation system, we choose four primary criteria that include cost, time, energy, and quality. Furthermore, to quantitatively measure the QoS, three subindicators of it are exacted from the evaluation information of CMS, which are pass rate, OTDR, and reliability respectively.

1. Cost (C): the execution cost of a CMS, including all the cost related to the machining process, handling, storage, etc.
2. Delivery time (DT): the expected date that providers deliver the tasks. Borrowing the philosophy of just-in-time (JIT), neither earliness nor tardiness should be avoided referring to the task's due date. The parameter $\Delta T = \max\{DT - dt, 0\}$ is defined as the tardiness, where dt denotes the due date.
3. Pass rate (PR): the pass rate of the finished product which can be derived from the historical production record.
4. *OTDR*: the possibility of delivering tasks on time according to the *DT*.
5. Reliability (R): the execution reliability of the machine.
6. Energy (E): the energy consumption of the machine; the electricity consumption in the service process is primarily concerned in this work.

The GRA-based evaluation method is presented to perform the service optimal selection, helping demanders find their satisfactory services. The whole procedure consists of following steps:

1. Generate the initial evaluation matrix

$$S=\begin{bmatrix} s_1^1 & \cdots & s_n^1 \\ \cdots & s_j^i & \cdots \\ s_1^m & \cdots & s_n^m \end{bmatrix}_{m\times n}$$

Let s_j^i denote the jth indicator of ith service, where $1 \le i \le m$, $1 \le j \le n$; m is the total number of candidate services, n is the number of indicators then $n = 6$.

2. Determine the optimal indicator sequence
The indicators are treated by one of the three types, one is the benefit-oriented indicator, that is, the larger the better; another is the cost-oriented indicator, that is, the smaller the better; the other is the nominal is the best.

Definition 5.1: $s_j^+ = \max\limits_{1\le i\le m} s_j^i, s_j^- = \min\limits_{1\le i\le m} s_j^i$

$$s_j^* = \begin{cases} s_j^+, & s_j^i \in I_b \\ s_j^-, & s_j^i \in I_c \\ s_j^o, & s_j^i \in I_o \end{cases}, \qquad j = 1,2,\ldots,n \tag{5.8}$$

Let s_j^o denote the target value of the jth indicator, where the I_b is the set of benefit-oriented indicators, I_c is the set of cost-oriented indicators, and I_o is the set of indicators with target values. Thus, the optimal sequence is achieved as: $S^* = (s_1^*, s_2^*, \ldots, s_n^*)$.

3. Normalize the evaluation matrix
For the benefit-oriented indicators, the initial sequence is normalized as follows:

$$\gamma_j^i = \frac{s_j^i - s_j^-}{s_j^+ - s_j^-}, \quad s_j^i \in I_b, j = 1,2,\ldots,n \tag{5.9}$$

For the cost-oriented indicators:

$$\gamma_j^i = \frac{s_j^+ - s_j^i}{s_j^+ - s_j^-}, \quad s_j^i \in I_c, j = 1,2,\ldots,n \tag{5.10}$$

For the indicators with the desired values:

$$\gamma_j^i = \begin{cases} \dfrac{s_j^i - s_j^o}{s_j^+ - s_j^o} & s_j^o < s_j^- \\ 1 - \dfrac{\left|s_j^o - s_j^i\right|}{\max(s_j^+ - s_j^o, s_j^o - s_j^-)} & s_j^- \le s_j^o \le s_j^+ \\ \dfrac{s_j^o - s_j^i}{s_j^o - s_j^-} & s_j^+ < s_j^o \end{cases} \tag{5.11}$$

Therefore, the initial evaluation matrix is revised as $S_N = [\gamma_j^i]_{m\times n}$. Accordingly, the optimal sequence is normalized as: $\gamma^* = (\gamma_1^*, \gamma_2^*, ..., \gamma_n^*)$.

4. Calculate the grey relational coefficient
In GRA, the parameter ξ_j^i presents the relational coefficient between s_j^i and s_j^*. It is calculated as shown in Eq. (5.12).

$$\xi_j^i = \frac{\min\limits_{1\le i\le m}\min\limits_{1\le j\le n}\left|\gamma_j^* - \gamma_j^i\right| + \rho \max\limits_{1\le i\le m}\max\limits_{1\le j\le n}\left|\gamma_j^* - \gamma_j^i\right|}{\left|\gamma_j^* - \gamma_j^i\right| + \rho \max\limits_{1\le i\le m}\max\limits_{1\le j\le n}\left|\gamma_j^* - \gamma_j^i\right|} \tag{5.12}$$

where ρ is the distinguishing coefficient, and it is typically taken as 0.5. Therefore, the relational coefficient matrix is derived as $E = [\xi_j^i]_{m\times n}$.

5. Calculate the grey relational degree
The victor $w = (\mu_1, \mu_2, ..., \mu_n)^T$ presents the weights of each indicator. Demanders customize the weight victor to meet their specific requirements. The grey relational degrees are calculated and the comprehensive evaluation matrix is obtained as:

$$R[x^i] = Ew \tag{5.13}$$

where x^i denotes relational degree between the ith candidate and the optimal sequence. According to the relational degrees, the candidates can be prioritized and the one with the highest value of relational degree can be considered the best service.

After evaluating the candidate sets, in which the CMSs are ranked in descending order in terms of their relational degrees, and top K_g services are selected from the total N_g candidates in $CMSCS_g$ to reduce the solution space when constructing the service compositions. Theoretically, there are total $\prod_{g=1}^{N} K_g$ compositions. The evaluation of service compositions is the same as that evaluates individual CMSs. The indicators for service composition are computed as shown in Table 5.1, where $DT(CMS^N)$ is the final DT of the service composition; and dt^N is the due date of the part-level task. After all the service compositions are evaluated, then the optimal CMS composition solution is generated.

TABLE 5.1 Evaluation Indicators for Service Composition

Evaluation indicator	Function
C	$C=\sum_{g=1}^{N} C(CMS_g)$
DT	$DT = DT(CMS^N)$ $\Delta T = \max\{DT(CMS^N) - dt^N, 0\}$
PR	$PR = \prod_{g=1}^{N} PR(CMS_g)$
$OTDR$	$OTDR = \prod_{g=1}^{N} OTDR(CMS_g)$
R	$R = \prod_{g=1}^{N} R(CMS_g)$
E	$E = \sum_{g=1}^{N} E(CMS_g)$

REFERENCES

[1] L. Da Xu, W. He, S. Li, Internet of things in industries: a survey, IEEE Trans. Ind. Inform. 10 (4) (2014) 2233–2243.

[2] G.Q. Huang, Y.F. Zhang, P.Y. Jiang, RFID-based wireless manufacturing for walking-worker assembly islands with fixed-position layouts, Robot. Comput. Integr. Manuf. 23 (4) (2007) 469–477.

[3] K. Chard, K. Bubendorfer, S. Caton, O.F. Rana, Social cloud computing: a vision for socially motivated resource sharing, IEEE Trans. Serv. Comput. 5 (4) (2012) 551–563.

[4] F. Tao, L. Zhang, V.C. Venkatesh, Y. Luo, Y. Cheng, Cloud manufacturing: a computing and service-oriented manufacturing model, Proc. Inst. Mech. Eng. B 225 (2011) 1969–1976.

[5] L. Ren, L. Zhang, F. Tao, C. Zhao, X. Chai, X. Zhao, Cloud manufacturing: from concept to practice, Enterp. Inf. Syst. 9 (2) (2015) 186–209.

[6] L. Ren, L. Zhang, L. Wang, F. Tao, X. Chai, Cloud manufacturing: key characteristics and applications, Int. J. Comput. Integr. Manuf. January (2014) 1–15.

[7] G. Škulj, R. Vrabicˇ, P. Butala, A. Sluga, Decentralised network architecture for cloud manufacturing, Int. J. Comput. Integr. Manuf. 3052 (March) (2015) 1–14.

[8] X. Xu, From cloud computing to cloud manufacturing, Robot. Comput. Integr. Manuf. 28 (1) (2012) 75–86.

[9] H.-H. Wu, A comparative study of using grey relational analysis in multiple attribute decision making problems, Qual. Eng. 15 (2) (2002) 209–217.

[10] F. Tao, Y. Zuo, L. Da Xu, L. Zhang, IoT-based intelligent perception and access of manufacturing resource toward cloud manufacturing, IEEE Trans. Ind. Inform. 10 (2) (2014) 1547–1557.

[11] T.C. Manufacturing, S. System, F. Tao, Y. Cheng, L. Da Xu, S. Member, L. Zhang, B.H. Li, CCIoT-CMfg: cloud computing and Internet of things, IEEE Trans. Ind. Inform. 10 (2) (2014) 1435–1442.

[12] O. Morariu, C. Morariu, T. Borangiu, Shop-floor resource virtualization layer with private cloud support, J. Intell. Manuf. 27 (2) (2014) 447–462.

[13] N. Liu, X. Li, Granulation-based resource classification in cloud Manufacturing, Proc. Inst. Mech. Eng. B 229 (7) (2015) 1–13.

[14] Y. Luo, L. Zhang, F. Tao, L. Ren, Y. Liu, Z. Zhang, A modeling and description method of multidimensional information for manufacturing capability in cloud manufacturing system, Int. J. Adv. Manuf. Technol. 69 (5–8) (2013) 961–975.

[15] W. Xu, J. Yu, Z. Zhou, Y. Xie, D.T. Pham, C. Ji, Dynamic modeling of manufacturing equipment capability using condition information in cloud manufacturing, J. Manuf. Sci. Eng. 137 (2015) 1–36.

[16] Y. Zhang, G. Zhang, Y. Liu, D. Hu, Research on services encapsulation and virtualization access model of machine for cloud manufacturing, J. Intell. Manuf. (2015).

[17] L. Guo, S. Wang, L. Kang, Y. Cao, Agent-based manufacturing service discovery method for cloud manufacturing, Int. J. Adv. Manuf. Technol. 81 (9–12) (2015) 2167–2181.

[18] H. Li, L. Zhang, R. Jiang, Study of manufacturing cloud service matching algorithm based on OWL-S, in: Control Decis. Conf. (2014 CCDC), 26th Chinese, 2014, pp. 4155–4160.

[19] B. Huang, C. Li, C. Yin, X. Zhao, Cloud manufacturing service platform for small- and medium-sized enterprises, Int. J. Adv. Manuf. Technol. 65 (9–12) (2013) 1261–1272.

[20] T. Song, H. Liu, C. Wei, C. Zhang, Common engines of cloud manufacturing service platform for SMES, Int. J. Adv. Manuf. Technol. 73 (1–4) (2014) 557–569.

[21] X.Wu, Q. Fei, K. Poon, Cloud manufacturing applications in semiconductor industry, in: Proc. Winter Simul. Conf. January, 2015 pp. 2376–2383.

[22] X. Yang, G. Shi, Z. Zhang, Collaboration of large equipment complete service under cloud manufacturing mode, Int. J. Prod. Res. 52 (February) (2014) 326–336.

[23] R.Y. Zhong, S. Lan, C. Xu, Q. Dai, G.Q. Huang, Visualization of RFID-enabled shopfloor logistics Big Data in cloud manufacturing, Int. J. Adv. Manuf. Technol. 84 (1–4) (2016) 5–16.

[24] S. Li, L. Da Xu, S. Zhao, The Internet of things: a survey, Inf. Syst. Front. 17 (2) (2015) 243–259.

[25] R.Y. Zhong, Q.Y. Dai, T. Qu, G.J. Hu, G.Q. Huang, RFID-enabled real-time manufacturing execution system for mass-customization production, Robot. Comput. Integr. Manuf. 29 (2) (2013) 283–292.

[26] Y. Zhang, G.Q. Huang, T. Qu, O. Ho, S. Sun, Agent-based smart objects management system for real-time ubiquitous manufacturing, Robot. Comput. Integr. Manuf. 27 (3) (2011) 538–549.

[27] G.Q. Huang, T. Qu, Y. Zhang, H.D. Yang, RFID- enabled product-service system for automotive part and accessory manufacturing alliances, Int. J. Prod. Res. 50 (14) (2012) 3821–3840.

[28] Z. Yang, P. Zhang, L. Chen, RFID-enabled indoor positioning method for a real-time manufacturing execution system using OS-ELM, Neurocomputing 174 (2014) 121–133.

[29] Y. Zhang, G. Zhang, J. Wang, S. Sun, S. Si, T. Yang, Real-time information capturing and integration framework of the Internet of manufacturing things, Int. J. Comput. Integr. Manuf. 3052 (December) (2014) 1–12.

[30] Y. Zhang, G.Q. Huang, S. Sun, T. Yang, Multi-agent based real-time production scheduling method for radio frequency identification enabled ubiquitous shopfloor environment, Comput. Ind. Eng. 76 (1) (2014) 89–97.

[31] H. Luo, J. Fang, G.Q. Huang, Real-time scheduling for hybrid flowshop in ubiquitous manufacturing environment, Comput. Ind. Eng. 84 (2015) 12–23.

[32] R.Y. Zhong, G.Q. Huang, S. Lan, Q.Y. Dai, T. Zhang, C. Xu, A two-level advanced production planning and scheduling model for RFID-enabled ubiquitous manufacturing, Adv. Eng. Inform. 29 (4) (2015) 799–812.
[33] Z.X. Guo, E.W.T. Ngai, C. Yang, X. Liang, An RFID-based intelligent decision support system architecture for production monitoring and scheduling in a distributed manufacturing environment, Int. J. Prod. Econ. 159 (2015) 16–28.
[34] I. Arkan, H. Van Landeghem, Evaluating the performance of a discrete manufacturing process using RFID: a case study, Robot. Comput. Integr. Manuf. 29 (6) (2013) 502–512.
[35] Y. Zhang, G. Zhang, W. Du, J. Wang, E. Ali, S. Sun, An optimization method for shopfloor material handling based on real-time and multi-source manufacturing data, Int. J. Prod. Econ. 165 (2015) 282–292.
[36] R.Y. Zhong, G.Q. Huang, S. Lan, Q.Y. Dai, X. Chen, T. Zhang, A big data approach for logistics trajectory discovery from RFID-enabled production data, Int. J. Prod. Econ. 165 (2015) 260–272.
[37] T. Qu, S.P. Lei, Z.Z. Wang, D.X. Nie, X. Chen, G.Q. Huang, IoT-based real-time production logistics synchronization system under smart cloud manufacturing, Int. J. Adv. Manuf. Technol. 84 (1–4) (2016) 147–164.
[38] K. Wang, Intelligent and integrated RFID (II-RFID) system for improving traceability in manufacturing, 2014, pp. 106–120.
[39] F. Tao, D. Zhao, Y. Hu, Z. Zhou, Resource service composition and its optimal-selection based on particle swarm optimization in manufacturing grid system, IEEE Trans. Ind. Inform. 4 (4) (2008) 315–327.
[40] F. Tao, Y. Laili, L. Xu, L. Zhang, FC-PACO-RM: a parallel method for service composition optimal-selection in cloud manufacturing system, IEEE Trans. Ind. Inform. 9 (4) (2013) 2023–2033.
[41] B. Huang, C. Li, F. Tao, A chaos control optimal algorithm for QoS-based service composition selection in cloud manufacturing system, Enterp. Inf. Syst. 8 (4) (2013) 445–463.
[42] W. Liu, B. Liu, D. Sun, Y. Li, G. Ma, Study on multi-task oriented services composition and optimisation with the 'Multi-Composition for Each Task' pattern in cloud manufacturing systems, Int. J. Comput. Integr. Manuf. 26 (8) (2013) 786–805.
[43] B. Liu, Z. Zhang, QoS-aware service composition for cloud manufacturing based on the optimal construction of synergistic elementary service groups, Int. J. Adv. Manuf. Technol. 2 (2016) 1–15.
[44] F. Xiang, Y. Hu, Y. Yu, H. Wu, QoS and energy consumption aware service composition and optimal-selection based on Pareto group leader algorithm in cloud manufacturing system, Cent. Eur. J. Oper. Res. 22 (4) (2014) 663–685.
[45] X. Xue, S. Wang, B. Lu, Manufacturing service composition method based on networked collaboration mode, J. Netw. Comput. Appl. 59 (1) (2015) 1–11.
[46] H. Zheng, Y. Feng, J. Tan, A fuzzy QoS-aware resource service selection considering design preference in cloud manufacturing system, Int. J. Adv. Manuf. Technol. 84 (1–4) (2016) 371–379.
[47] F. Tao, D. Zhao, H. Yefa, Z. Zhou, Correlation-aware resource service composition and optimal-selection in manufacturing grid, Eur. J. Oper. Res. 201 (1) (2010) 129–143.
[48] H. Jin, X. Yao, Y. Chen, Correlation-aware QoS modeling and manufacturing cloud service composition, J. Intell. Manuf. (2015).
[49] Y. Cao, S. Wang, L. Kang, Y. Gao, A TQCS-based service selection and scheduling strategy in cloud manufacturing, Int. J. Adv. Manuf. Technol. 82 (1–4) (2016) 235–251.
[50] J. Lartigau, X. Xu, L. Nie, D. Zhan, Cloud manufacturing service composition based on QoS with geo-perspective transportation using an improved Artificial Bee Colony optimisation algorithm, Int. J. Prod. Res. 53 (14) (2015) 4380–4404.

[51] F. Xiang, G.Z. Jiang, L.L. Xu, N.X. Wang, The case-library method for service composition and optimal selection of big manufacturing data in cloud manufacturing system, Int. J. Adv. Manuf. Technol. 84 (1–4) (2016) 59–70.

[52] Y. Lu, Q. Shao, C. Singh, X. Xu, Ontology for manufacturing resources in a cloud environment, Int. J. Manuf. Res. 9 (4) (2014) 448–469.

[53] N. Srinivasan, M. Paolucci, K. Sycara, An efficient algorithm for OWL-S based semantic search in UDDI, Semant, Web Serv. Web (2005) 96–110.

[54] Uddi Org, UDDI Technical White Paper, *Int. Bus.*, p. www.uddi.org/pubs/Iru UDDI Technical White Paper. p. 2000.

[55] M. Paolucci, T.b Kawamura, T.R. Payne, K. Sycara, Semantic matching of web services capabilities, Lect. Notes Comput. Sci. (including Subser. Lect. Notes Artif. Intell. Lect. Notes Bioinformatics), vol. 2342 LNCS, 2002, pp. 333–347.

Chapter 6

IoT-Enabled Smart Trolley

6.1 INTRODUCTION

With the rapid development of industrial wireless network and auto-ID technologies [e.g., radio frequency identification (RFID), Bluetooth, Wi-Fi], many enterprises adopt these advanced technologies to implement real-time traceability, visibility, and interoperability in improving the performance of shop-floor planning, execution, and control [1], and huge streams of manufacturing information has been produced during diverse manufacturing processes.

In the shop floor, material handling plays an important role in improving the production efficiency. It has been widely considered by academia and industry. To improve the transport efficiency and reduce the transport time, Herrmann et al. [2] designed the material flow networks. A heuristic method has been put forward by Anwar et al. [3] to solve the simultaneous scheduling problem of material handling transporters in the production of complex assembled product. Lee et al. [4] provide dispatching strategies for the rail-guided vehicle scheduling problem in a flexible manufacturing system. Khayat et al. [5] propose an integrated formulation to solve the combined production and material handling scheduling problems. Asef-Vaziri et al. [6] develop the heuristic procedures to minimize the total loaded and empty vehicle trip distances. Boonprasurt et al. [7] present an optimal model for the vehicle routing problem with manual materials handling.

Recently, real-time and multisource information are more accessible and ubiquitous since Internet of things (IoT) technologies (e.g., RFID or barcode) have been extended to the manufacturing environment [8]. Real-time information streams are captured by the sensor network in manufacturing environment, which imposes new challenges on the optimization model of material handling system. Lee et al. [9] state that material handling will be more efficient based on the massive real-time information.

Though researchers have made great progress in improving the material handling, major challenges still exist in how to utilize real-time and multisource manufacturing information to make better decision for shop-floor material handling (SMH). They are summarized as follows:

1. Generally, move tasks are centrally allocated to trolleys according to a given objective, for example, the minimum total transport time. However, during the

Optimization of Manufacturing Systems Using the Internet of Things
http://dx.doi.org/10.1016/B978-0-12-809910-0.00006-8

material handling execution, the deviations between the plan and execution are often caused by unpredictable exceptions. The more the move tasks, the bigger the deviations. Then, these deviations will lead to further serious production exceptions. Therefore, how to develop a real-time information-based allocation strategy of material handling tasks to reduce or avoid deviations should be considered.
2. Trolleys in the shop floor can sense different manufacturing resources (e.g., material, pallet, and operator) and actively request moves by attached intelligent and auto-ID devices. In addition, their real-time status could be timely tracked and traced. The delivery efficiency could also be enhanced by implementing real-time information-driven intelligent navigation during material handling process. Therefore, how to use auto-ID devices and information technologies to enable the trolleys have the capability of active sensing and intelligence plays an important role in improving the material handling in manufacturing shop floor.
3. Traditional material handling methods rarely take the combination of move tasks into account. However, for the purpose of low carbon, it's necessary to adopt a combination method of move tasks to improve the material handling efficiency and decrease the transport cost. Therefore, how to combine the different move tasks according to the priority of tasks, maximum load, and volume of the trolleys is another issue in implementing green material handling in the manufacturing shop floor.

To address the aforementioned challenges, a novel optimization model for SMH is proposed in this chapter. It is composed of three key points. The first one is an active allocation strategy of move tasks. The second point is the IoT-enabled smart trolleys with the capability of active sensing and self-decision. The third point is the combination optimization method of move tasks to decrease the transport cost and energy consumption. The presented optimization method will provide a new paradigm for manufacturing enterprises to implement real-time information-driven SMH.

6.2 RELATED WORKS

There are two streams of literature that are relevant to this research. They are material handling and real-time data capturing in manufacturing field.

6.2.1 Material Handling

Material handling allows matching vendor supply with customer demand, smoothing demand for seasonal products, consolidating products, customizing product, or packaging and arranging distribution activities [10]. Many optimization models have been proposed to improve the material handling. Nazzal et al. [11] develop a closed queuing network approach to analyze the multivehicle material handling systems. Lau et al. [12] present an artificial immune system (AIS)-based model to provide an effective methodology to coordinate and control multiagent systems. Lau and Woo [13] introduce a dynamic routing strategy for determining the optimal route for material flow under a distributed agent-based material handling system. Tuzkaya et al. [14] apply an integrated fuzzy multicriteria decision-making methodology to solve the material handling equipment selection problem. To minimize the machine operation, material handling, and machine setup costs and maximize the machine utilization, Mahdavi et al. [15] designed a simulation-based optimization for controlling operation allocation and material handling equipment selection. To minimize the sum of the expected long-run average transport job waiting cost, Lin et al. [16] propose a Markov decision model–based dynamic vehicle allocation control for automated material handling system (AMHS) in semiconductor manufacturing. Dai et al. [17] explores the economic feasibility of a flexible material handling system using free-ranging automated guided vehicles (AGV) with a local positioning system (LPS) for the apparel industry. Poon et al. [18] develop a RFID-GA-based warehouse resource allocation system to solve stochastic production material demand problems. To effectively analyze and evaluate the performances of closed-loop AMHS with shortcut and blocking in semiconductor wafer fabrication system, a modified Markov chain model (MMCM) has been proposed [19]. Drießel and Mönch [20] suggest an extension of the shifting bottleneck heuristic for complex job shops that takes the operations of AMHS into account. By a combination of the concept of compromise solution and grey relational model, a new fuzzy grey multicriteria decision-making method is presented to deal with the evaluation and selection problems of material handling equipment under the condition of uncertain information [21]. Chung [22] develops a new heuristic method based on a stochastic approach to estimate the arrival times of transportation jobs to their final destinations for automated material handling in an LCD fabrication facility. Because of the need for steadiness and stability in the automated manufacturing systems, a biobjective stochastic programming model is used to evaluate material handling systems with automated guide vehicles [23]. A variant of particle swarm optimization (PSO) with a self-learning strategy is provided for vehicle routing problem of multiple products with material handling in multiple cross-docks [24]. Choe et al. [25] investigates how cognitive automation and mechanical automaton in the material handling system affect manufacturing flexibility. Their simulation program shows that cognitive automation has a very critical effect on manufacturing flexibility in the material handling system. Hadi-Vencheh et al. [26] propose a new hybrid fuzzy multicriteria decision-making model for solving the equipment selection problem in material handling system. The standard deviation, variance, and the downside risk of the cost distribution are investigated as the risk measures of material handling system by Mital et al. [27].

6.2.2 Real-Time Data Capturing in Manufacturing Field

As a novel paradigm, IoT is rapidly gaining ground in the scenario of modern wireless telecommunications [28]. By enabling instant identification and automatic information transfer, IoT technologies (e.g., RFID, auto ID) provide visibility of a dynamically moving object in a constantly changing environment [29]. Experts have done lots of researches about applications of IoT technologies in manufacturing process. RFID has been widely used in supporting the logistics management on manufacturing shop floors where production resources attached with RFID facilities are converted into smart manufacturing objects (SMOs) which are able to sense, interact, and reason to create a ubiquitous environment [30]. Lee et al. [31] present an RFID-based resource allocation system (RFID-RAS), integrating RFID technology and fuzzy logic concept for achieving better resource allocation with particular reference to garment manufacturing. Zhang et al. [32] provide an innovative all-in-one Smart Gateway technology for capturing real-time production data from various manufacturing resources attached to different types of RFID/auto-ID devices. Zhang et al. [33] use the RFID-enabled reconfigurable manufacturing sources to achieve real-time agent-based workflow management. Tu et al. [34] propose an agent-based distributed production control framework with UHF RFID technology to help firms adapt to such a dynamic and agile manufacturing environment. Guo et al. [35] design RFID-based intelligent decision support system architecture to handle production monitoring and scheduling in a distributed manufacturing environment. To monitor and control dynamic production flows and also to improve the traceability and visibility of mass customization manufacturing processes, Chen et al. [36] developed an agent-based manufacturing control and coordination (AMCC) system, an agent-based framework using ontology and RFID technology. Huang et al. [37] developed RFID-based real-time collaborative manufacturing shop-floor service platforms to address automotive manufacturing standards and practices within automotive parts and accessories manufacturers. A study has been made to show the capability of a RFID-based information system in the international distribution process of a car manufacturer [38]. Zhou and Piramuthu [39] consider RFID tags and their applications from a recycling/remanufacturing perspective and propose a novel framework to assist such process based on item-level information visibility and instantaneous tracking/tracing ability enabled by RFID. An intelligent and real-time multiobjective decision-making model is developed to provide timely and effective solutions for multiobjective production planning problem by integrating RFID technology with intelligent optimization techniques [40]. Zhong et al. [41] propose an RFID-enabled real-time advanced production planning and scheduling shell (RAPShell, in short) to coordinate different decision makers across production processes. To address important logistics operations aspects, Mejjaouli and Babiceanu use an integrated RFID-sensor network system to detect the condition of perishable products as they are moved downstream the supply chain before undesired total loss of products occurs [42]. Tang et al. [43] propose a value-driven uncertainty-aware data-processing method that considers RFID detection reliability, timeliness, and the throughput of an assembly line to characterize the potential benefits of RFID implementation in a mixed-model assembly system. Fan et al. [44] implement a research on the impact of RFID technology adoption on supply chain decisions with shrinkage and misplacement problems in the IoT. Oliveira et al. [45] propose a model for logistics management based on geofencing algorithms and radio-frequency technology to improve services, reduce costs, and ensure the safety in cargo transportation.

Aforementioned researches make significant contribution to solving the material handling problems. However, they mainly focus on a traditional manufacturing environment. Extending IoT technologies to manufacturing environment will bring new decision strategies for SHM in many perspectives. For example, real-time decision models will be promoted by the massive real-time and multisource manufacturing information. Therefore, several research issues should be further studied in the IoT-based manufacturing environment. The first issue is about a new material handling strategy in the IoT-based manufacturing environment. The second issue is how to develop IoT-based trolleys to execute intelligent material handling. The third issue is how to efficiently combine the different move tasks for the IoT-based trolleys according to their real-time status. To address these issues, a novel optimization model is designed to implement real-time information-driven SMH in the IoT-based manufacturing environment.

6.3 REAL-TIME INFORMATION ENABLED MATERIAL HANDLING STRATEGY

The material handling referred in this chapter mainly focuses on a discrete manufacturing and fixed position assembly environment. The presented optimization model for SMH aims to enable distribution resources to have interactive ability through extending IoT technologies to the material handling process, and implement the real-time information-driven allocation of move tasks.

Fig. 6.1 describes the central material handling strategy. The central material handling strategy is generally adopted by traditional manufacturing environment. In this strategy, move tasks are centrally allocated to trolleys by the material handling system, and there is no interaction between trolleys and other distribution resources. Moreover, the decision model of this strategy rarely takes the real-time information of distribution resources into account. Deviations between the plan and execution are often caused by unpredictable exceptions. The more the move tasks, the bigger the deviations. In addition, due to the increasing move tasks and trolleys, the computational complexity becomes higher and higher.

Fig. 6.2 describes the real-time information driven active material handling strategy. In the real-time information driven active material handling strategy, each trolley could actively request move tasks according to its

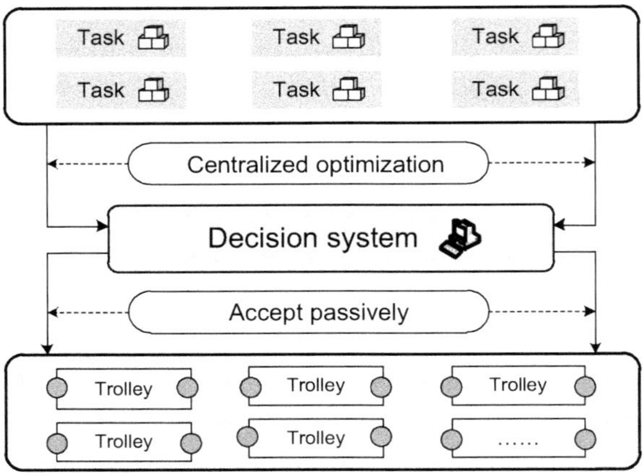

FIGURE 6.1 Central material handling strategy.

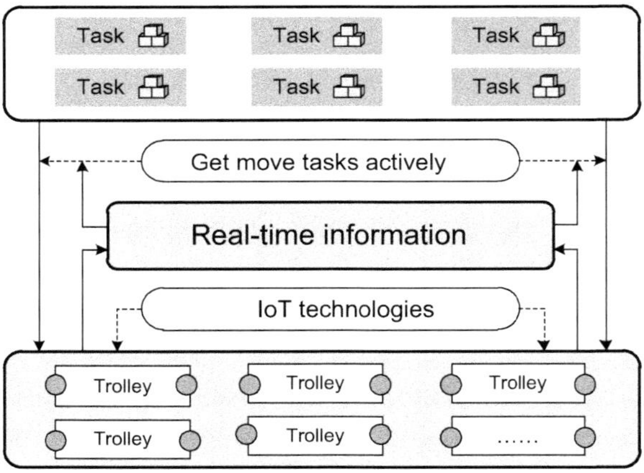

FIGURE 6.2 Active material handling strategy.

real-time status. Because IoT technologies are extended to the manufacturing environment, the real-time information of trolleys could be easily captured, and continuous interacts between trolleys and move tasks server could also be achieved. Based on the captured real-time information, trolleys will get the most suitable move tasks when they are idle. After trolleys complete the allocated move tasks, they will automatically send their current status and request the move tasks again. This procedure will repeat until all the move tasks are finished.

The presented real-time information driven active material handling strategy has following advantages. First, each idle trolley can get the optimal move tasks at any time. Second, because only one trolley requests move tasks at each time, the complexity of this strategy is stable. Third, since the move task allocation is real-time information driven and the allocation process has been only started for the idle trolleys, the deviations between plan and execution in central material handling strategy can be largely removed in the active material handling strategy.

6.4 OVERALL ARCHITECTURE OF OPTIMIZATION MODEL FOR SMH

As shown in Fig. 6.3, a conceptual architecture of the optimization model for SMH is designed based on the proposed real-time information driven active material handling strategy. It consists of three modules, namely IoT-enabled smart trolley, real-time information exchange, and combination optimization method for move tasks.

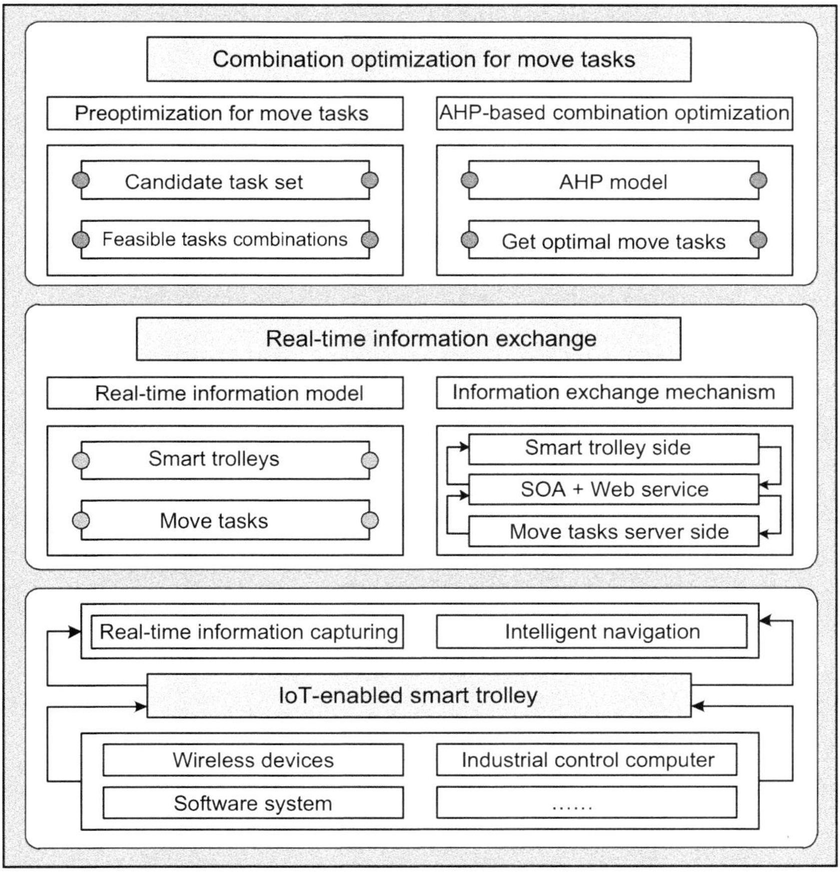

FIGURE 6.3 Overall architecture of optimization model for SMH.

IoT-enabled smart trolley module is the core part to achieve the real-time information driven active material handling strategy. It is responsible for enabling the trolleys to get the capability of active perception and dynamic interaction by adopting IoT technologies. As a middle section, real-time information exchange module provides a mechanism for the real-time information exchanging between distributed trolleys and move task server. Combination optimization for move tasks module aims to implement green SMH through combining the move tasks according to priority of move tasks, maximum load, and volume of the trolleys.

6.5 IoT-ENABLED SMART TROLLEY

The capturing and usage of real-time and multisource information of trolleys play an important role in the proposed active material handling strategy. How to use advanced IoT technologies (e.g., RFID, web services, and workflow) to make traditional trolleys smart will be described in detail in this section.

As said before, IoT-enabled smart trolley module aims to enable trolleys to get the capability of active capturing, interaction, and self-decision. Overall solution of IoT-enabled smart trolley is shown in Fig. 6.4. Three parts are included in the solution. They are real-time information capturing and encapsulation, real-time information exchange, workflow-based real-time navigation. The functions of these parts are described as follows.

6.5.1 Real-Time Information Capturing and Encapsulation

This part is responsible for making trolleys get the capability of active perception. As shown in the middle of Fig. 6.4, some hardware devices (e.g., industrial

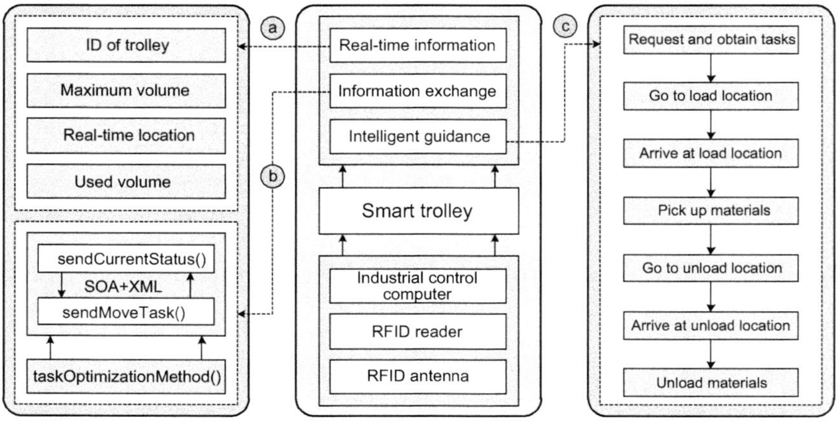

FIGURE 6.4 Overall solution of the proposed smart trolley.

control computer, RFID reader and antennas) are attached to the trolley. They are employed to timely capture the real-time information. The employed hardware devices mainly include industrial control computer, RFID reader and antennas. The industrial control computer is used to store temporary real-time information and develop some software. RFID reader and antennas are employed to capture real-time information of manufacturing resources (e.g., pallets, operators, and key work-in-progress) which are attached RFID tags. Their detailed information is described as follows:

- Industrial control computer
  The main parameters of the employed industrial control computer in this chapter include its model, capacity of its memory, and capacity of its hard disk. Its model is PCA-6007LV. The capacity of its memory is 1G. The capacity of its hard disk is 160G. The functions of industrial control computer can be categorized into three aspects. First, it can connect RFID readers and receive the captured real-time data. Second, it can send the real-time status of the trolley to information server and get the assigned move tasks from the information server. Third, it can provide the real-time navigation information for the operators.
- RFID reader
  The main parameters of the employed RFID reader in this chapter refer to its model, working frequency, RF power, interface style, number of antenna interfaces, and its reading distance. Its model is XAFD6132C. Its working frequency varies from 902 $\sim$ 928 MHz. Its RF power is under 30 DBM. It supports the interface styles such as RS232, RS485, TCPIP. It has four antenna interfaces. Its longest reading distance is 2 m. The functions of RFID reader can be categorized into two aspects. First, it can connect the three antennas for sensing the operator, material items, and position data. Second, it can transmit the captured real-time data to industrial control computer through RS232.
- Antenna
  The main parameters of the employed antenna in this chapter refer to its type, its Gain, its VSWR, its F/B ratio, and impedance. Its type is UHF. Its Gain is 16dBi. Its VSWR is d1.5. Its F/B ratio is more than 25 dB. Its impedance is 50 Ω. Three antennas are deployed at the trolley side. They are used to sense the real-time data of the tags attached to different manufacturing resources such as operator, material items, and position data.

As shown in the upper-left of Fig. 6.4, to manage and exchange real-time information during material handling process, an information model is constructed to describe the real-time information of trolley. Four information nodes are included in the constructed model. They are ID of trolley, maximum volume of trolley, real-time location of trolley, and used volume of trolley.

TABLE 6.1 The Defined Notations for the Information Model of Trolley

VID_i	Code of trolley *i*
$V^i_{\max}$	Maximum volume of trolley *i*
CL_i	Real-time location of trolley *i*
V^i_u	Used volume of trolley *i*

The matrix V is used to store these nodes. The defined notations are listed in Table 6.1.

$$V = \begin{bmatrix} VID_1 & V^1_{max} & CL_1 & V^1_u \\ VID_2 & V^2_{max} & CL_2 & V^2_u \\ VID_3 & V^3_{max} & CL_3 & V^3_u \\ \cdots & \cdots & \cdots & \cdots \\ VID_i & V^i_{max} & CL_i & V^i_u \end{bmatrix}$$

6.5.2 Real-Time Information Exchange

This part is responsible for completing the real-time information exchange between move tasks server and trolleys. The worklogic of real-time information exchanging mechanism is shown in the lower-left of Fig. 6.4. Service-oriented architecture (SOA) is employed in this part to send and receive the real-time information during material handling. First, an XML-based schema about the real-time information of trolley will be formed. It contains the information such as trolley ID, maximum volume of trolley, current location, and current operator of trolley. Second, the formed XML-based instance will be sent to move task server by the sendCurrentStatus() method. Then, the web service of move task server side can receive the real-time information of the requested trolley, and the taskOptimizationMethod() will be invoked to get the optimal move task according to the real-time information of trolley and move tasks. Finally, the sendMoveTask() method will be used to send the optimal result to the trolley side. The detailed information of the mentioned optimization method will be illustrated in Section 6.6.

6.5.3 Workflow-Based Real-Time Guidance

This part is responsible for providing trolleys and operators with the self-decision capability and real-time information-driven guidance. To achieve this aim, workflow is introduced to this part. The key processes during material handling are shown in the right of Fig. 6.4. They are as follows: (1) request and obtain task, (2) go to load location, (3) arrive at the load location, (4) pick up materials, (5) go to unload location, (6) arrive at the unload location, and (7) unload materials. The operations (2) to (4) or operations (5) to (7) is repeated when multimove tasks are assigned to the trolley.

During the material handling process, the real-time guidance will be graphically shown in the screen. This guidance could help the operators efficiently and correctively complete the material handling.

6.6 TWO-STAGE COMBINATION OPTIMIZATION METHOD FOR MOVE TASKS

To combine and optimize the material handling tasks based on the real-time information, a two-stage combination optimization method for move tasks is designed in this section. By analyzing the sensed real-time information and integrating material handling requirements, the designed method could assign optimal combination of move tasks to the idle trolley. As the name suggests, this method contains two stages. The first stage aims to get a candidate move tasks set from the whole move tasks. The second stage aims to get an optimal combination of move tasks from the formed candidate move tasks set. The proposed method could decrease the complexity of obtaining optimal move tasks and promote the real-time decision during material handling.

6.6.1 Real-Time Information Models of Move Tasks

The information model of move tasks is essential to describe the optimization method. Assume that there are N tasks in the move task pool. The information of each move task contains following nodes: ID of move task, from and to locations of move task, due time of move task, priority of move task, material index number of move task, and material information (i.e., ID of material, the quantity of material and volume of material) matched with its index number. The matrix N stores the information model of N move tasks. The defined notations are listed in Table 6.2.

$$N = \begin{bmatrix} TID_1 & FL_1 & TL_1 & D_1 & P_1 & IID_1 \\ TID_2 & FL_2 & TL_2 & D_2 & P_2 & IID_2 \\ \cdots & \cdots & \cdots & \cdots & \cdots & \cdots \\ TID_j & FL_j & TL_j & D_j & P_j & IID_j \\ \cdots & \cdots & \cdots & \cdots & \cdots & \cdots \\ TID_N & FL_N & TL_N & D_N & P_N & IID_N \end{bmatrix}$$

The information model of materials contains following nodes: ID of material, name of material, the quantity of material, and unit volume of each material.

TABLE 6.2 The Defined Notations for the Information Model of Move Task

TID_j	Code of task *j*
FL_j	From-location of task *j*
TL_j	To-location of task *j*
D_j	Due time of task *j*
P_j	Priority of task *j*
IID_j	Material index number of task *j*
$ICode_{jk}$	Code of material *k*
$Name_{jk}$	Name of material *k*
Q_{jk}	The quantity of material *k*
V_{jk}	Unit volume of material *k*

Matrix *Wj* is used to store the material information model of task *j*. The defined notations are listed in Table 6.3.

$$W_j = \begin{bmatrix} ICode_{j1} & Name_{j1} & Q_{j1} & V_{j1} \\ ICode_{j2} & Name_{j2} & Q_{j2} & V_{j2} \\ ICode_{j3} & Name_{j3} & Q_{j3} & V_{j3} \\ \cdots & \cdots & \cdots & \cdots \\ \cdots & \cdots & \cdots & \cdots \\ ICode_{jk} & Name_{jk} & Q_{jk} & V_{jk} \end{bmatrix}$$

TABLE 6.3 The Defined Notations for Combination Optimization

c	One task combination
z	Number of material kind in one task
m	Number of tasks in one tasks combination
L_c	Distribution distance of task combination *c*
P_c	Priority of task combination *c*
U_c	Used volume of task combination *c*
w_L	Weights of *L* in function $f(L_c,P_c,U_c)$
w_P	Weights of *P* in function $f(L_c,P_c,U_c)$
w_U	Weights of *U* in function $f(L_c,P_c,U_c)$
P_0, L_0, U_0	Used to unify the dimensions of *P,L,U*
P_{cj}	Priority of task *j* in task combination *c*

6.6.2 Preoptimization for Candidate Tasks Set

There are lots of move tasks in the move tasks server side. To reduce the calculating complexity, it is necessary to implement the preoptimization for candidate task set to select better tasks for further combination optimization. The preoptimization for candidate tasks set could be executed according to the priorities of move tasks, which are defined by their due time. In general, the earlier the due time, the higher the priority.

The following manner could be used to get the candidate tasks set. Select q tasks from the move tasks server. The priority values of these selected tasks are higher than other tasks in the tasks pool. If tasks have the same priority values during the selecting process, select the task having early due time. Matrix q is used to store the information of candidate task set.

$$q = \begin{bmatrix} TID_{q.1} & FL_{q.1} & TL_{q.1} & D_{q.1} & P_{q.1} & IID_{q.1} \\ TID_{q.2} & FL_{q.2} & TL_{q.2} & D_{q.2} & P_{q.2} & IID_{q.2} \\ TID_{q.3} & FL_{q.3} & TL_{q.3} & D_{q.3} & P_{q.3} & IID_{q.3} \\ \cdots & \cdots & \cdots & \cdots & \cdots & \cdots \\ \cdots & \cdots & \cdots & \cdots & \cdots & \cdots \\ TID_{q.q} & FL_{q.q} & TL_{q.q} & D_{q.q} & P_{q.q} & IID_{q.q} \end{bmatrix}$$

After getting the candidate tasks set, we can form the feasible tasks combinations. A feasible tasks combination may be composed of one or more move tasks which are selected from the candidate tasks set. The volume of any feasible tasks combination must not exceed the maximum volume of current trolley. It is the principle that should be obeyed during forming the feasible tasks combinations.

6.6.3 AHP-Based Combination Optimization

As a decision making tool, AHP has been widely used to make decisions in different engineering fields [46]. Due to the promising features of AHP in solving engineering problems, an AHP-based combination method is used to get the optimal move tasks combination from candidate tasks set. To implement the green material handling, the transport distance of trolley, the priorities of move tasks and the used volume of trolley are considered in the presented AHP-based combination method.

During the combination optimization process, the AHP is used to identify the weight coefficients for the transport distance, priority, and the used volume of the trolley in obtaining the optimal tasks combination. The AHP model of obtaining the optimal tasks combination is shown in Fig. 6.5. In the AHP model, material handling cost, material handling time, and the material handling quality belong to criterion layer; the transport distance, priority of tasks combination and the used volume of trolley belong to the project layer.

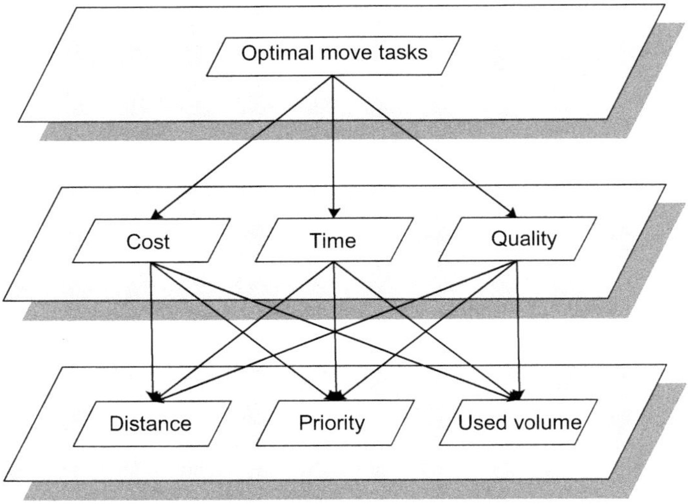

FIGURE 6.5 **The AHP model of the optimal combination for the real-time handling tasks.**

To get the reasonable weight coefficients, it is necessary to adopt the paired comparison method to obtain the judgment matrix (A), which could be represented as follow.

$$A = (a_{ij})_{n\times n} = \begin{bmatrix} a_{11} & a_{12} & \cdots & a_{1n} \\ a_{21} & a_{22} & \cdots & a_{2n} \\ \vdots & \vdots & \vdots & \vdots \\ a_{n1} & a_{n2} & \cdots & a_{nn} \end{bmatrix}$$

Then, the weight coefficients could be calculated as formula (6.1)

$$w_i = \frac{\sqrt[n]{\prod_j a_{ij}}}{\sum \sqrt[n]{\prod_j a_{ij}}} \text{and} \sum_{i=1}^{n} w_i = 1, W = (w_1, w_2, \cdots, w_n)T \tag{6.1}$$

The consistency test is introduced to verify the effectiveness of the obtained weight coefficients. It contains following steps.

First, establish the feature equation of the matrix A according to equation (6.2) and getting the maximum value $\lambda_{\max}$ according to Eq. (6.3). Second, calculate the coincident indicator I_{CI} according to equation (6.4) and attain the average random coincident indicator $I_{RI}(n)$ corresponding to the variable n. Finally, make the consistency test according to Eq. (6.5). If the variable I_{CR} satisfies the following constraint condition $I_{CR} < 0.1$, the judgment matrix is acceptable. Then, the weight coefficients for the transport distance, priority of tasks candidate, and the used volume of the trolley in getting the optimal tasks combination are identified as the vector: $(w_L, w_P, w_U)^T$.

$$A\lambda_{\max} = \lambda_{\max} W \tag{6.2}$$

$$\lambda_{\max} = \frac{1}{n}\sum_{i}\left(\frac{(AW)_i}{w_i}\right) \tag{6.3}$$

$$I_{CI} = \frac{\lambda_{\max} - n}{n-1} \tag{6.4}$$

$$I_{CR} = \frac{I_{CI}}{I_{RI}} \tag{6.5}$$

For better understanding, the notations are defined as seen in Table 6.3.

The AHP-based combination optimization method can be implemented after the weights of transport distance, priority of move tasks and used volume are obtained based on aforementioned steps. As shown in Fig. 6.6, the procedure of implementing the combination optimization method contains following steps:

Step 1: Construct the objective function. The objective of this problem is to minimize the weighted transport distance L_c and maximize the weighted priority of tasks combination P_c and used volume of trolley U_c. So the objective function could be stated as formula (6.6). Here, P_0, L_0, U_0 are used to unify the dimensions of P, L, U.

$$\max f(L_c, P_c, U_c) = w_L L_0 / L_c + w_P P_c / P_0 + w_U U_c / U_0 \tag{6.6}$$

Step 2: Get the weights w_L, w_P, w_U according to the mentioned AHP steps.
Step 3: Calculate the transport distance of completing task combination c. Take the shortest distance of current trolley completing the task combination as L_c.
Step 4: Calculate the priority of the tasks combination c. Take the summation of priorities of all tasks in the tasks combination as the priority of the tasks combination. P_{cj} is defined as the priority of task j in tasks combination c.
Step 5: Calculate used volume of task combination c.
Step 6: Calculate P_0, L_0, U_0. As defined before, P_0, L_0, U_0 are used to unify the dimensions of P, L, U. Here, we take the average value of P, L, U of all the feasible task combinations as P_0, L_0, U_0.
Step 7: Calculate the value of $f(L_c, P_c, U_c)$ according to the formula (6.6).

The bigger the value of $f(L_c, P_c, U_c)$ is, the better the tasks combination matches with current trolley. The information of the tasks combination with biggest value of $f(L_c, P_c, U_c)$ will be sent to the trolley side. Then, the trolley will implement the material handling according to the received information.

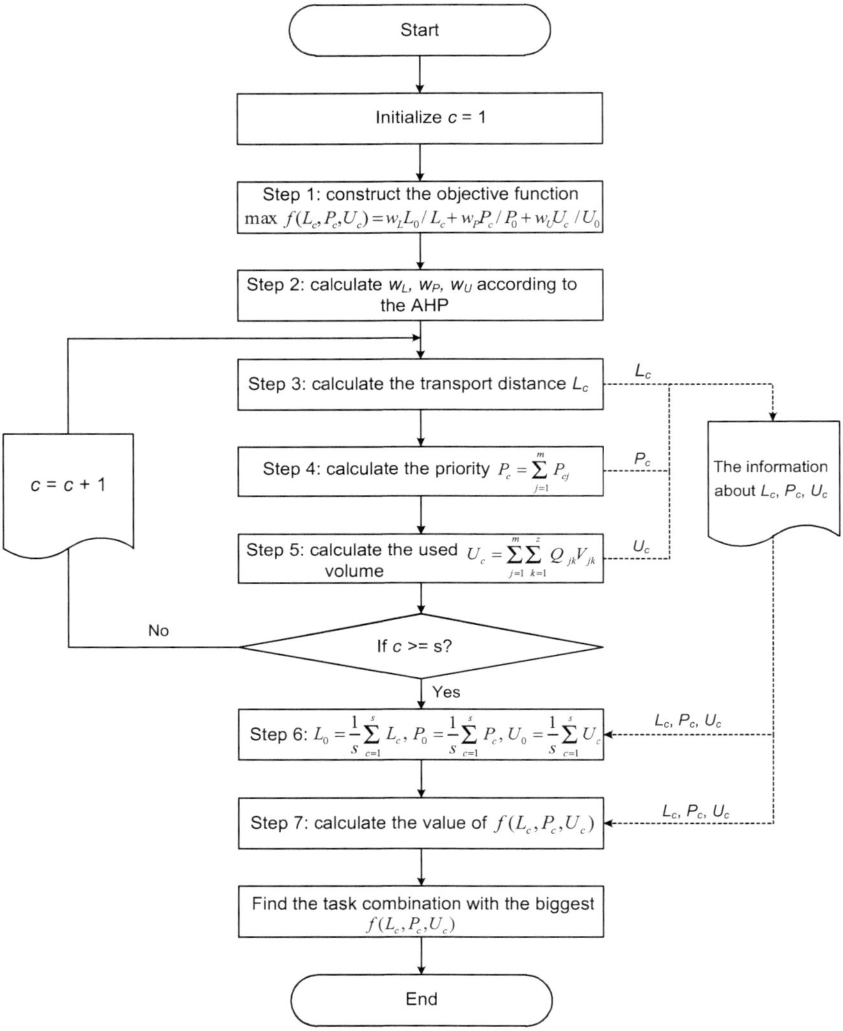

FIGURE 6.6 The procedure of implementing the combination optimization method.

REFERENCES

[1] G.Q. Huang, P.K. Wright, S.T. Newman, Wireless manufacturing: a literature review, recent developments, and case studies, Int. J. Comput. Integr. Manuf. 22 (7) (2009) 579–594.

[2] J.W. Herrmann, G. Ioannou, I. Minis, R. Nagi, J.M. Proth, Design of material flow networks in manufacturing facilities, J. Manuf. Syst. 14 (4) (1995) 277–289.

[3] M.F. Anwar, R. Nagi, Integrated scheduling of material handling and manufacturing activities for just-in-time production of complex assemblies, Int. J. Prod. Res. 36 (3) (1998) 653–681.

[4] J. Lee, Dispatching rail-guided vehicles and scheduling jobs in a flexible manufacturing system, Int. J. Prod. Res. 37 (1) (1999) 111–123.
[5] G.E. Khayat, A. Langevin, D. Riopel, Integrated production and material handling scheduling using mathematical programming and constraint programming, Eur. J. Oper. Res. 175 (3) (2006) 1818–1832.
[6] A. Asef-Vaziri, G. Laporte, R. Ortiz, Exact and heuristic procedures for the material handling circular flow path design problem, Eur. J. Oper. Res. 176 (2) (2007) 707–726.
[7] P. Boonprasurt, S. Nanthavanij, Optimal fleet size, delivery routes, and workforce assignments for the vehicle routing problem with manual materials handling, Int. J. Ind. Eng. Theory Appl. Pract. 19 (6) (2012) 252–263.
[8] Y. Zhang, G. Zhang, J. Wang, S. Sun, S. Si, T. Yang, Real-time information capturing and integration framework of the internet of manufacturing things, Int. J. Comput. Integr. Manuf. 28 (8) (2015) 811–822.
[9] J. Lee, E. Lapira, B. Bagheri, H. an Kao, Recent advances and trends in predictive manufacturing systems in big data environment, Manuf. Lett. 1 (1) (2013) 38–41.
[10] R. Manzini, Y. Bozer, S. Heragu, Decision models for the design, optimization and management of warehousing and material handling systems, Int. J. Prod. Econ. 170 (2015) 711–716.
[11] D. Nazzal, A closed queueing network approach to analyzing multi-vehicle material handling systems, IIE Trans. 43 (10) (2011) 721–738.
[12] H.Y.K. Lau, V.W.K. Wong, A.K.S. Ng, A cooperative control model for multiagent-based material handling systems, Expert Syst. Appl. 36 (1) (2009) 233–247.
[13] H.Y.K. Lau, S.O. Woo, An agent-based dynamic routing strategy for automated material handling systems, Int. J. Comput. Integr. Manuf. 21 (3) (2008) 269–288.
[14] G. Tuzkaya, B. Gülsün, C. Kahraman, D. Özgen, An integrated fuzzy multi-criteria decision making methodology for material handling equipment selection problem and an application, Expert Syst. Appl. 37 (4) (2010) 2853–2863.
[15] I. Mahdavi, B. Shirazi, N. Sahebjamnia, Development of a simulation-based optimisation for controlling operation allocation and material handling equipment selection in FMS, Int. J. Prod. Res. 49 (23) (2011) 6981–7005.
[16] J.T. Lin, C.H. Wu, C.W. Huang, Dynamic vehicle allocation control for automated material handling system in semiconductor manufacturing, Comput. Oper. Res. 40 (10) (2013) 2329–2339.
[17] J.B. Dai, N.K.S. Lee, Economic feasibility analysis of flexible material handling systems: a case study in the apparel industry, Int. J. Prod. Econ. 136 (1) (2012) 28–36.
[18] T.C. Poon, K.L. Choy, C.K. Cheng, S.I. Lao, H.Y. Lam, Effective selection and allocation of material handling equipment for stochastic production material demand problems using genetic algorithm, Expert Syst. Appl. 38 (10) (2011) 12497–12505.
[19] J. Zhang, W. Qin, L.H. Wu, A performance analytical model of automated material handling system for semiconductor wafer fabrication system, Int. J. Prod. Res. 54 (6) (2015) 1–20.
[20] R. Drießel, L. Mönch, An integrated scheduling and material-handling approach for complex job shops: a computational study, Int. J. Prod. Res. 50 (20) (2012) 5966–5985.
[21] S.M. Mousavi, B. Vahdani, R. Tavakkoli-Moghaddam, N. Tajik, Soft computing based on a fuzzy grey group compromise solution approach with an application to the selection problem of material handling equipment, Int. J. Comput. Integr. Manuf. 27 (6) (2014) 547–569.
[22] J. Chung, Estimating arrival times of transportation jobs for automated material handling in LCD fabrication facilities, J. Manuf. Syst. 35 (2015) 112–119.
[23] M. Tavana, H. Fazlollahtabar, R. Hassanzadeh, A bi-objective stochastic programming model for optimising automated material handling systems with reliability considerations, Int. J. Prod. Res. 52 (19) (2014) 5597–5610.

[24] M. Chen, Y. Hsiao, R. Himadeep Reddy, M.K. Tiwari, The Self-Learning Particle Swarm Optimization approach for routing pickup and delivery of multiple products with material handling in multiple cross-docks, Transp. Res. E 91 (2016) 208–226.
[25] P. Choe, J.D. Tew, S. Tong, Effect of cognitive automation in a material handling system on manufacturing flexibility, Int. J. Prod. Econ. 170 (2013) 891–899.
[26] A. Hadi-Vencheh, A. Mohamadghasemi, A new hybrid fuzzy multi-criteria decision making model for solving the material handling equipment selection problem, Int. J. Comput. Integr. Manuf. 28 (5) (2015) 534–550.
[27] P. Mital, M. Goetschalckx, E. Huang, Robust material handling system design with standard deviation, variance and downside risk as risk measures, Int. J. Prod. Econ. 170 (2015) 815–824.
[28] L. Atzori, A. Iera, G. Morabito, The Internet of things: a survey, Comput. Netw. 54 (15) (2010) 2787–2805.
[29] J. Kim, C.-S. Ok, S. Kumara, S.-T. Yee, A market-based approach for dynamic vehicle deployment planning using radio frequency identification (RFID) information, Int. J. Prod. Econ. 128 (1) (2010) 235–247.
[30] R.Y. Zhong, G.Q. Huang, S. Lan, Q.Y. Dai, X. Chen, T. Zhang, A big data approach for logistics trajectory discovery from RFID-enabled production data, Int. J. Prod. Econ. 165 (2015) 260–272.
[31] C.K.H. Lee, K.L. Choy, G.T.S. Ho, K.M.Y. Law, A RFID-based Resource Allocation System for garment manufacturing, Expert Syst. Appl. 40 (2) (2013) 784–799.
[32] Y. Zhang, T. Qu, O.K. Ho, G.Q. Huang, Agent-based Smart Gateway for RFID-enabled real-time wireless manufacturing, Int. J. Prod. Res. 49 (5) (2011) 1337–1352.
[33] Y. Zhang, G.Q. Huang, T. Qu, O. Ho, Agent-based workflow management for RFID-enabled real-time reconfigurable manufacturing, Int. J. Comput. Integr. Manuf. 23 (2) (2010) 101–112.
[34] M. Tu, J.H. Lin, R.S. Chen, K.Y. Chen, J.S. Jwo, Agent-based control framework for mass customization manufacturing With UHF RFID technology, IEEE Syst. J. 3 (3) (2009) 343–359.
[35] Z.X. Guo, E.W.T. Ngai, C. Yang, X. Liang, An RFID-based intelligent decision support system architecture for production monitoring and scheduling in a distributed manufacturing environment, Int. J. Prod. Econ. 159 (2015) 16–28.
[36] R.S. Chen, M. (Arthur) Tu, Development of an agent-based system for manufacturing control and coordination with ontology and RFID technology, Expert Syst. Appl. 36 (4) (2009) 7581–7593.
[37] G.Q. Huang, T. Qu, Y. Zhang, H.D. Yang, RFID- enabled product-service system for automotive part and accessory manufacturing alliances, Int. J. Prod. Res. 50 (14) (2012) 3821–3840.
[38] S. Herrmann, H. Rogers, M. Gebhard, E. Hartmann, Co-creating value in the automotive supply chain: an RFID application for processing finished vehicles, Prod. Plan. Control 26 (12) (2015) 981–993.
[39] W. Zhou, S. Piramuthu, Remanufacturing with RFID item-level information: optimization, waste reduction and quality improvement, Int. J. Prod. Econ. 145 (2) (2013) 647–657.
[40] W.K. Wong, Z.X. Guo, S.Y.S. Leung, Intelligent multi-objective decision-making model with RFID technology for production planning, Int. J. Prod. Econ. 147 (Part C) (2014) 647–658.
[41] R.Y. Zhong, Z. Li, L.Y. Pang, Y. Pan, T. Qu, G.Q. Huang, RFID-enabled real-time advanced planning and scheduling shell for production decision making, Int. J. Comput. Integr. Manuf. 26 (7) (2013) 649–662.
[42] S. Mejjaouli, R.F. Babiceanu, RFID-wireless sensor networks integration: decision models and optimization of logistics systems operations, J. Manuf. Syst. 35 (2015) 234–245.

[43] L. Tang, H. Cao, L. Zheng, N. Huang, Value-driven uncertainty-aware data processing for an RFID-enabled mixed-model assembly line, Int. J. Prod. Econ. 165 (2015) 273–281.

[44] T. Fan, F. Tao, S. Deng, S. Li, Impact of RFID technology on supply chain decisions with inventory inaccuracies, Int. J. Prod Econ. 159 (2015) 117–125.

[45] R.R. Oliveira, I.M.G. Cardoso, J.L.V. Barbosa, C.A. Da Costa, M.P. Prado, An intelligent model for logistics management based on geofencing algorithms and RFID technology, Expert Syst. Appl. 42 (15–16) (2015) 6082–6097.

[46] O.S. Vaidya, S. Kumar, Analytic hierarchy process: an overview of applications, Eur. J. Oper. Res. 169 (1.) (2006) 1–29.

Chapter 7

Real-Time Key Production Performances Analysis Method

7.1 INTRODUCTION

With the rapid growth of smart sensing technologies, it is possible to fulfill the smart manufacturing with the ability of dynamically sensing, optimal scheduling, and real-time control [1]. The innovative era of Internet of things (IoT) is generated as a consequence of the widespread use of sensor technologies, such as RFID, Wi-Fi, GSM, and so on. It denotes uniquely distinguishable things and their virtual depictions under an Internet-alike framework [2–4]. By applying the IoT devices to traditional manufacturing field, real-time and multisource data (RMD) from manufacturing activities can become more available and pervasive, as discussed in Chapter 3.

Amounts of distributed manufacturing data are obtained by the sensors, most of which are often meaningless and cannot be used by the upper level managers directly. However, the production anomalies often disturb the normal manufacturing activities. The key production performance analysis (KPPA) is the base to find the anomalies. If the production performance (PP) cannot be on time, the anomaly cannot be found and managed in time, which will lead to the aggravation of process disturbance. Therefore, it is important for the enterprises managers to improve their production management pattern with advanced technologies and models to fulfill the real-time KPPA, so that the managers can make optimal decisions based on real-time production status and the anomaly can be detected and eliminated dynamically.

In general, Petri nets (PNs) are recognized to be dominant in process modeling of discrete event system (DES), both graphically and mathematically [5,6]. Decision tree (DT) is a supervised knowledge acquisition technique, which can extract the rules from either expert knowledge or historical data, and it is a widely used data-driven approach in decision support and rule obtainment [7,8]. Hence, the PNs and DTs are used as our KPPA tools.

By combining IoT technologies, PNs, and DTs, this work develops an all-in-one real-time key production performance analysis method (PPAM). The aim of this chapter is to answer the following two great challenges: (1) how to build up an events-driven real-time key PPAM to process the huge real-time data

Optimization of Manufacturing Systems Using the Internet of Things
http://dx.doi.org/10.1016/B978-0-12-809910-0.00007-X

captured by distributed auto-ID devices to meaningful and value-added PP information? (2) How to efficiently analyze the PP, so that the real-time anomaly can be extracted timely and the causes of the anomaly can be diagnosed?

The rest of this chapter is arranged as follows. Section 7.2 reviews the related works. Section 7.3 describes the overall architecture of real-time KPPAM. Section 7.4 gives the event hierarchy of multilevel event, while the extraction of critical event (CrE) is presented in Section 7.5. Section 7.6 gives the real-time production anomaly analysis method.

7.2 RELATED WORKS

Three categories of literature are relevant to this research. They are real-time production monitoring technique, real-time production key performance indicators (KPIs) analysis, and real-time production anomaly analysis.

7.2.1 Real-Time Production Monitoring Technique

In the area of real-time production monitoring technique, Guo et al. proposed an RFID-based smart decision support system architecture for production monitoring and scheduling in a distributed manufacturing environment. The RFID, cloud technology, and intelligent techniques are used in the architecture [9]. Chai et al. presented a cooperative manufacturing execution systems; it uses real-time information to realize lean production [10]. By applying the concept of cloud computing into manufacturing field, Wang and Xu provided an interoperable production mode [11]. Wang et al. developed a framework, that is, Wise-ShopFloor (web-based-integrated sensor-driven e-ShopFloor), for real-time monitoring and remote control of networked CNC machines [12]. Chen et al. proposed an RFID-based enterprise application integration framework for real-time management of dynamic manufacturing processes [13]. Chongwatpol and Sharda proposed a creative information visibility-based scheduling (VBS) rule that utilized information generated from the real-time traceability systems for tracking work in processes (WIPs), parts and components, and raw materials to adjust production schedules [14]. Arkan and Van proposed an RFID-based real-time location system (RTLS) solution for obtaining multiitem WIP visibility within a company [15]. Bevilacqua et al. presented a case study for implementation of an RFID system in a furniture industry involved in the fashion sector; the case company is an Italian enterprise which is a leader in the furniture industry [16].

7.2.2 Real-Time Production KPIs Analysis

In the production KPIs analysis aspect, Georgiadis and Michaloudis merged discrete event simulation and continuous dynamics in a unified hybrid simulation environment. By coordinating the simulations run in real time, the results can be associated with information from the shop-floor frontline [17]. Zang et al. applied an event-processing method for management systems using RFID, where the framework, information structures, optimization manners, and algorithms are discussed in detail [18]. Considering the difference businesses in different companies, Huang et al. presented an agent-based workflow management method [19] and Zhang et al. proposed a smart objects management system for RFID-enabled real-time reconfigurable manufacturing [20]. Kumaraguru et al. proposed an approach to integrate real-time analytics with continuous performance management, which can form the base for further research on fulfilling likely interoperability issues and essential standardization efforts to support improvement of a system [21]. Blaha et al. designed a paradigm that enables real-time monitoring using cognitive models and its implementation was demonstrated with a fatigue-sensitive task [22]. Li proposed an aggregation method by overlapping decomposition approach to approximate the production rate for systems with rework loops; the systems were decomposed into four serial production lines, with machines modified to accommodate the interactions with other machines and buffers [23]. Colledani and Tolio designed a new approximate analytical method to estimate the quality and productivity performance measures of asynchronous production lines, thus the associations between the quality control and system production logistics performance can be captured [24]. Lao and Liu integrated data envelopment analysis (DEA) and geographic information systems (GIS) to evaluate the performance of bus lines within a public transit system, considering both the operations and operational environment [25]. Lin et al. proposed a three-phase method to measure the performance of a highly value-added footwear manufacturing system considering reworking actions, in which the system consists of multiple production lines [26]. Using RFID-enabled shop-floor production data, Zhong et al. proposed a data mining approach (DT-based model) to estimate realistic standard operation time and unknown dispatching rules [27].

The modeling of manufacturing activities plays an important role in real-time PPA. Lv et al. developed a novel RFID-based CPN analysis strategy, where the behaviors of the dynamic production status can be described through colored tokens [28]. To model the dynamic behavior of manufacturing systems, Wu and Zhou offered an intelligent token PN (ITPN) for modeling and control of reconfigurable production systems, where colored tokens are extended as macrotokens so that they can represent job instances with real-time awareness about system status just like smart cards in real life [29]. Zhang et al. used the hierarchy timed color Petri net (HTCPN) method to construct the real-time production events, where the events are displayed with a multilevel event model [30]. Başak and Albayrak presented a PN-based decision system modeling method for real-time scheduling and control of flexible automotive manufacturing systems [31]. Wang et al. proposed a new colored PN model for real-time scheduling of multiprocessor system on chip platform [32]. Similarly, a model that described the reconfiguring process of a manufacturing system was developed by adopting colored timed object-oriented PNs [33]. Ha and Suh designed PWF-nets based on timed colored PNs and proposed a method of organizing PWF-nets that are composed of workflow patterns, so that the difficulty in workflow management because of the uncertain and dynamic characteristics can be tackled [34].

7.2.3 Real-Time Production Anomaly Analysis

To analysis production anomalies, two main steps need to be taken, that is, the anomaly extraction and anomaly cause diagnosis. In the area of anomaly extraction, Evans et al. presented a distinct experience-based decision support system, which uses factual information of historical decision to compute the confidence factors [35]. Zarei et al. presented an intelligent method based on artificial neural networks (ANNs) to identify bearing defects of induction motors [36]. Cabral presented a PN approach for online diagnosis of DESs modeled by finite state automata; the method is based on the construction of a Petri net diagnoser (PND) which is constructed in polynomial time and requires less memory than other methods proposed in the literature [37]. Cabasino et al. proposed a method for online diagnosis of DESs based on labeled PNs [38]. He et al. proposed a DT learning-based model for bivariate process mean shift monitoring and fault identification [39]. Maier et al. presented new model-based approaches for testing and diagnosis of automation systems; the models are not created manually but learned automatically by monitoring the plant behavior [40]. Similarly, Faltinski et al. presented a new model-based approach for the prediction of energy consumption in production plants in order to detect anomalies. Hybrid timed automaton models of the supervised production plant are generated and executed in parallel to the system so that the anomalies can be detected automatically [41]. Yin et al. considered the real-time implementation of fault-tolerant control systems with performance optimization [42].

In the anomaly cause diagnosis aspect, Chioua et al. designed a top-down approach for root cause analysis using KPIs; the differences between bottom-up and top-down methods for root cause analysis are studied in the paper [43]. Wang et al. presented a graphic modeling approach, that is, fault diagnosis method based on fuzzy reasoning spiking neural P systems (FDSNP), for power transmission networks [44]. Groenewald and Aldrich proposed a root cause analysis method for processing fault situations on an industrial concentrator circuit by use of causality maps and extreme learning machines; the maps are combined with two approaches based on the use of extreme learning machines [45]. A DBN-based framework for root cause reasoning was proposed to deal with abnormal situation by Hu et al., and fault propagation behavior of process system is studied [46]. Alaeddini and Dogan designed a root cause analysis in statistical process control based on Bayesian networks (BNs); the cause and effect relationship among chart patterns, process information, and possible root causes/assignable causes are considered in the BNs [47]. An improved PLS (IPLS) approach is presented to cope with the problems for fault diagnosis related to KPI of the underlying process encountered by the standard approach [48].

Li et al. designed a new causality analysis index based on dynamic time warping to determine the causal direction between pairs of faulty variables [49].

7.2.4 Research Gap

Up to now, comprehensive research on the real-time production PAM has been done. However, less research effort is paid to the following issues.

1. Amounts of primitive data are captured after the multiple sensors are attached to the traditional manufacturing resources. However, most of the data are meaningless. There is a wide gap between the primitive data and the key PP information. It is necessary to build up an event-driven real-time KPPA model and procedure to process the huge real-time data captured by distributed auto-ID devices to meaningful and value-added manufacturing information.
2. Production anomalies are inevitable in real-time processes and the anomalies analyses are essential to ensure the normal manufacturing activities. In general, the anomalies are often found and diagnosed by the experts, thus the process for anomalies detection are always subjective. How to dynamically provide the upper level applications with the quantitative and qualitative information of anomalies persuasively and objectively is an important issue.

7.3 OVERALL ARCHITECTURE OF REAL-TIME PRODUCTION PERFORMANCE ANALYSIS MODEL

In order to meet the wide information gap between the physical manufacturing systems and enterprise information systems (EISs), this chapter proposes a real-time KPPAM. The IoT technologies, PNs, and DTs are combined to develop an efficient real-time KPPA architecture. The overall architecture of the proposed model is presented in Fig. 7.1. The KPPAM consists of three main modules, namely, configuration of smart sensors, CrE-based information extracting process, and real-time key production anomaly analysis. They are explained next.

7.3.1 Configuration of Smart Sensors

The goal of this module is to construct a smart manufacturing environment based on IoT technologies, thus the information bridge between physical manufacturing systems and upper-level information systems can be established. The IoT technologies, such as RFID, wireless, etc., are applied to the traditional manufacturing things based on the information requirements. Besides, the optimal algorithms are used to decide the configuration pattern, so that a smart manufacturing environment can be established in an optimal manner. At last, the amounts of manufacturing data can be timely obtained by upper-level applications. More details can be found in Chapter 3.

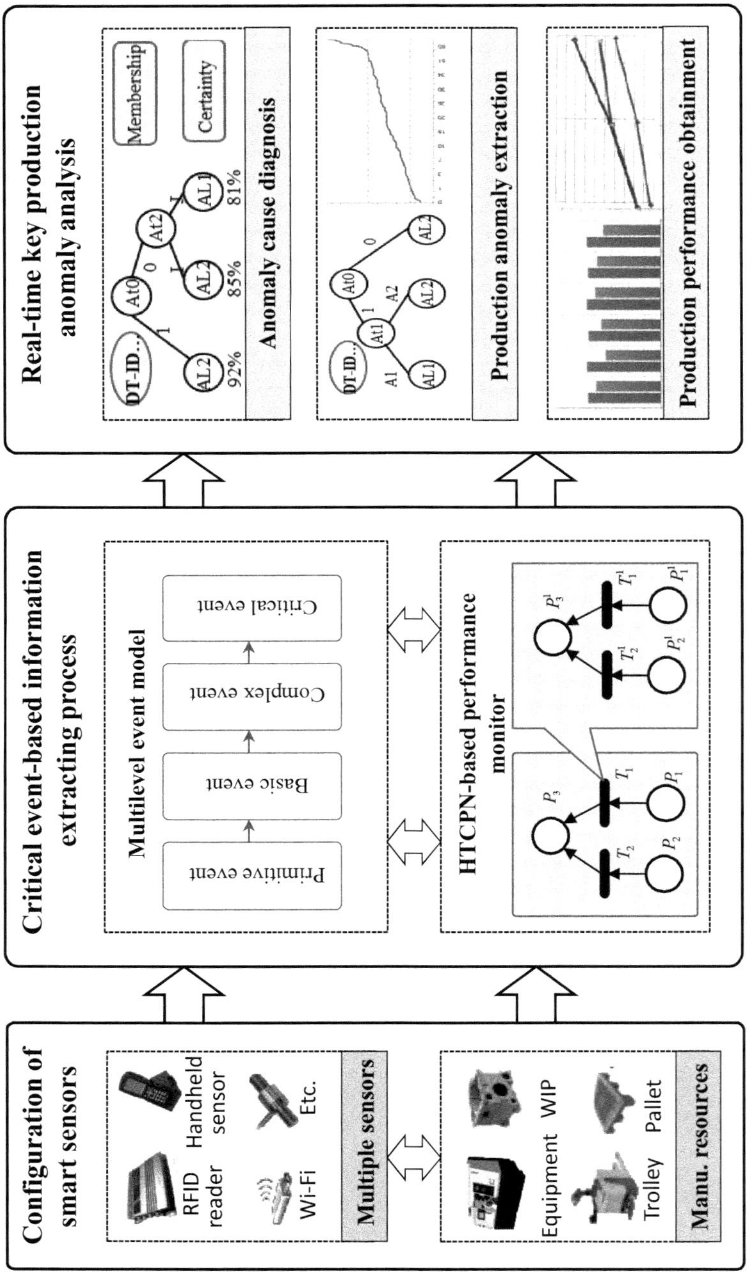

FIGURE 7.1 Overall architecture of real-time production performance analysis model.

7.3.2 Critical Event–Based Information Extracting Process

This module aims to model the dynamic behavior of the manufacturing system and process the large number of raw manufacturing data captured to CrEs of KPPA. As shown in the middle of Fig. 7.1, multilevel event model are used to analyze real-time PP. Four levels of events are established in this model, namely primitive events (PEs), basic events (BEs), complex events (CEs), and CrEs. The extracting processes are fulfilled by using HTCPN technologies, which will obtain the up-level events by continuously reading and integrating the primitive data. More details will be presented in Sections 7.4 and 7.5.

7.3.3 Real-Time Key Production Anomaly Analysis

Evaluating the real-time key PP condition is an important step in improving production efficiency. After the CrEs are obtained, the PPs are going to be evaluated to find the anomalies. First, the quantitative information of PP are obtained from the lower-level modules, and the information of the related factors are also called. Second, each kind of key PP are assessed to find the anomalies, respectively. The classic decision tree classifier, namely, C4.5, is used for the construction of anomaly extraction rules due to its ability to analyze continuous attributes. At last, the corresponding anomaly cause analysis rules are called to acquire the reasons as soon as an anomaly is extracted. Since most of the captured information of manufacturing resources has fuzzy attributes, the Fuzzy Interactive Dichotomizer 3 (Fuzzy-ID3) algorithm is used to obtain anomalies cause diagnosis rules. More details are given in Section 7.6.

7.4 THE EVENT HIERARCHY OF CRITICAL EVENT

The real-time production activities are complicated and interconnected, thereby the complexity of modeling and simulation is increased. The precise event model is the foundation of the KPPAM. As shown in Fig. 7.2, a four-level event hierarchy is discussed in this chapter. The detailed information of event at each level is shown as follows.

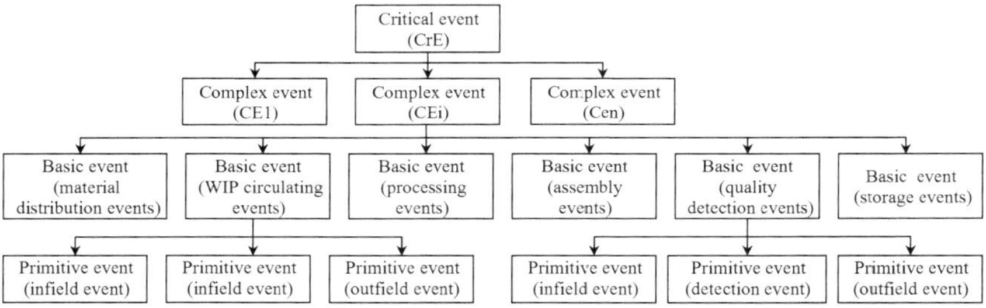

FIGURE 7.2 The event hierarchy of multilevel event.

PEs are raw events captured by the IoT devices. Due to the reading characteristic of high-speed and automatic reading of the IoT devices, the PEs are often obtained in a large volume. BEs are resource level events, which are formed by the aggregation of qualified raw data. Readers are referred to Chapter 3 for more details about PE and BE.

CEs are cell level events, for example, the process of a whole production line from material released to production offline. In general, the BEs denote the time and space status, and are connected with each other, so that they can be used to acquire the status of CE.

> *Definition 7.1*: CEs can be represented as $CE = (CE.ID, Attri, Context, T)$ where, *CE.ID* denotes the unique ID; *Attri* denotes the attributes, such as the event elements; *Context* describes the context information, such as material.ID, process.ID, the relationships among subevents, etc.; *T* is the time when the event occurs.

CrEs are highest level events, whose state changes have a large effect on the production management. Various enterprise information systems are concerned about different sides of PP, therefore, the CrEs have different meanings for them. For example, the change of total production cost is the CrE for the financial managers, while the real-time progress is the cared events for project managers.

> *Definition 7.2*: CrEs can be represented as $CrE = (CrE.ID, Attri, Context, T)$ where, *CrE.ID* is the unique ID, *Attri* denotes the attributes of the event, *Context* describes the context information, *T* denotes occur time.

7.5 HTCPN-BASED CRITICAL EVENT ANALYSIS

In order to acquire the CrEs, the HTCPN is used to model the constraint relationships and sequential relationships of the set of operations according to the multilevel events. Instead of modeling the static process, the HTCPN aims to model dynamic and hierarchical information of the manufacturing resources.

7.5.1 Basic Concepts of HTCPN

HTCPN is a graphical modeling language with a well-defined semantics, which has been proved to be a suitable tool to model dynamic manufacturing system. HTCPN, as a normal extension of the classic PN, have three main advantages for production modeling: (1) HTCPN can decrease the dimension of traditional

PN in a great extent; (2) HTCPN can model deterministic firing durations, which helps to obtain exact performance estimates; and (3) HTCPN can use a unified net to model similar components/subsystems, and the nets can be involved repeatedly when modeling a complex system.

Definition 7.3: An HTCPN is an 8-tuple $N = \langle P, T, C, I, O, G, D, M \rangle$
where, $P = \{P_t, P_m\}$ is the set of places, P_t and P_m are traditional and macroplaces, respectively. The places are used to represent resources status of the system; $T = \{T_i, T_t, T_s, T_m\}$ is the set of transitions, T_i, T_t, T_s, and T_m are immediate, timed, random, and macrotransitions, respectively. The macrotransitions are replaceable transitions that can be substituted by detailed sub-HTCPN systems. The transitions are used to represent the activity; C represents the color mapping from $P \cup T$ to W, an element of $C(s)$ is named as a color of s and $C(s)$ is the color set of s, s is the attribute of P or T. The colored tokens are used to represent particular type of product; the tokens can carry colored attributes, for example, time, quantity, and so on; $I(O)$ denotes the forward (backward) incidence matrix of $P \times T$, where $I(P, T)$ is a mapping from $C(P) \times C(T)$ to $N = \{0, 1, 2, \ldots\}$ and $O(P, T)$ is a mapping from $C(T) \times C(P)$ to $N = \{0, 1, 2, \ldots\}$. I/O functions are used to represent the relationship between transitions and places; G denotes the guard function, which is used to describe the trigger condition for each transition T; D represents the time duration of timed transition "T_t" or random transition "T_s." Once a transition in T_t or T_s is enabled, it cannot fire until D units of time are passed; M denotes a marking representing the number of tokens in P. M_0 represents the initial marking.

7.5.2 HTCPN Model Construction

To model the multilevel event, mainly five steps are considered to build an HTCPN.

Step 1: Summarize the outline of the production process in shop floor, and then construct the PN model for the CrE, where all the elements and constraints are considered into the PN.

Step 2: Further construct the sub-PNs for the upper-level HTCPN. Since the upper-level events are integrated by the lower-level events, the macrotransitions in the upper-level events can be replaced by substitutions which can be described in detail by sub-PNs.

Step 3: Repeat Step 2 until each bottom element is described by a PE captured by the sensor, which means that no further extension can be executed.

Step 4: Construct the contacts among the PNs at different levels, so that the relationships among the sub-PNs are corresponding with the entire multilevel model.

Step 5: Establish the connection between the HTCPN and the manufacturing resources. As a result, the HTCPN can model the dynamic behavior of the real-time manufacturing system.

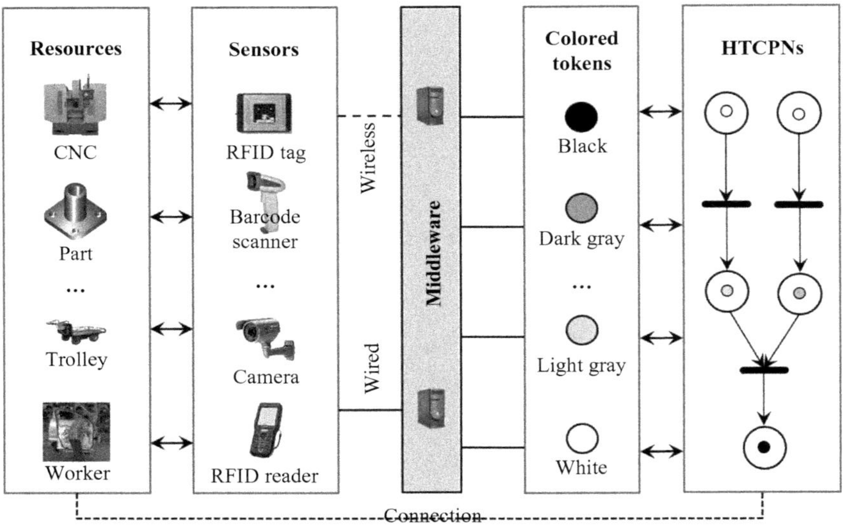

FIGURE 7.3 **Connection between the HTCPN and manufacturing resources.**

7.5.3 Connection Between HTCPN and Manufacturing Resources

As shown in Fig. 7.3, the connection between the established HTCPN and manufacturing resources is fulfilled by a middleware. On one hand, the manufacturing resources are monitored by the multiple sensors, and the real-time information are captured and uploaded to the middleware on time. On the other hand, the marking of the HTCPN alters based on the dynamic information acquired from the middleware. Thus, the tokens in the HTCPNs can be connected with the manufacturing resources. Once the status of a manufacturing resource alters, the color of the token will be altered immediately.

Fig. 7.4 shows a case for the PN-based production PAM. In the existing marking, four kinds of colored tokens are presented in the PN: two materials M_{01} (dark gray) in Place P_0, one material M_{02} (light gray) in Place P_1, and one Part M_0 (black) in Place P_2. Each token carries the related manufacturing information by the color. For example, the black token (M_0) carries the status information with five components (MID, WID, PQ, GT, PT): MID is the material ID, WID is the workstation ID, PQ is the present quality, GT is the getting time, and PT is the present time. If the manufacturing process works normally, the status of the token will alter following the logic of PN. For example, the aforementioned two dark gray tokens and one light gray token will be integrated to obtain a black token, that is to say, the materials are successfully manufactured. Nevertheless, if something goes erroneous in the process, the output token will alter to another color to display the state variation.

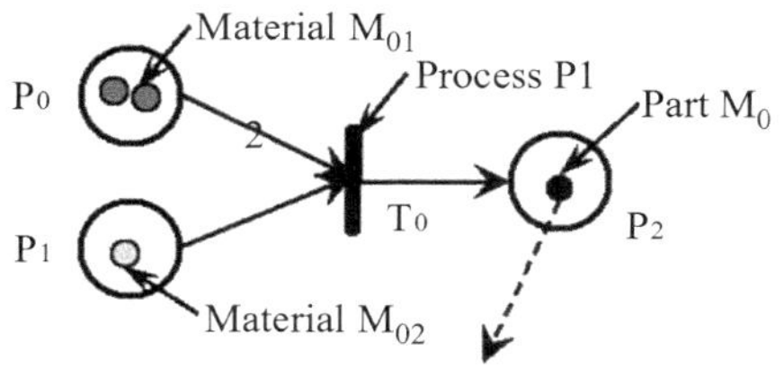

FIGURE 7.4 A case for PN-based production performance model.

7.5.4 Production Performance Extraction

After the HTCPN model is established and the tokens are linked with the manufacturing resources, the model can change its markings according to the real-time condition. Thus, the performance of the PN can reflect the real-time PP. In order to obtain the PP easily, three actions are used to analyze the manufacturing processes.

1. Observe intermediate marking for the real-time status of inventory and WIP. For that the tokens are connected with the resources in shop-floor frontline, the dynamic status can be reflected by the marking of the PNs.
2. Set additional places to record the firing status of the transitions, for example, the trigger frequency and time. Given that the transitions denote the process activity and the firing status records the process data (throughput, process quality, manufacturing time, etc.), the PP related to the process data can be obtained by setting extra places.
3. Analyze the performance of PNs to measure the PP related to overall production, for example, production cycle time, average time taken for jobs to wait in the queue, queue length, and so on.

From previously mentioned steps, the PP can be acquired after the simulation, and then the simulation report can be obtained, which will be uploaded to the upper-level applications to provide the performance information.

7.6 REAL-TIME PRODUCTION ANOMALY DIAGNOSIS

DT is one of the most popular machine learning techniques [7,50]. It can be used to establish the relationships between a large volume of input candidate attributes and an output attribute. This section discusses a DT-based anomaly extraction and cause diagnosis method for classification in a new manufacturing environment by learning from the historical cases.

Three main steps are essential in the classification. First, when a newly observed case comes, the related manufacturing information (PP and resource condition) of the historical cases is called, and the tree builder is triggered to construct the DTs for the production anomaly extraction and cause diagnosis. Then, the PP is assessed to find the anomaly according to the new rules from the DTs. At last, the anomaly cases will be further diagnosed to discover the causes of anomalies so that the managers can tackle the anomalies as soon as possible. The key elements in the DT-based production anomaly analysis are described as follows.

7.6.1 New Cases

In order to describe a new PP analysis problem simply and directly, the newly obtained information is represented as a new case (C).

$$C = \{P_t, R\} \tag{7.1}$$

where $P_t = \{P_{t1}, \ldots, P_{ti}, \ldots, P_{tn}\}$ is a set of attributes for PP, $P_{ti}(1 \leq i \leq n)$ is the ith kind of attribute, for example, the quantitative value. $R = \{R_1, \ldots, R_j, \ldots, R_m\}$ is a set of attributes for manufacturing resources. $R_j(1 \leq j \leq m)$ is the attribute of the jth kind of resource; they can be either crisp or fuzzy values.

7.6.2 Historical Cases

A new production anomaly analysis event is evaluated based on anomaly analysis rules. It is significant to feedback the exception analysis, so that experiences are learned by knowledge mined from historical cases. Thereby, the possibility of the success of classification for new anomalies can be increased.

The historical anomaly case can be denoted as

$$H = \{P_t, R, D\} \tag{7.2}$$

where, D is the serious degree of production anomaly, four levels of degrees are considered, that is, red, yellow, blue, and green.

7.6.3 Decision Variables

To evaluate production anomaly, multiple decision variables are involved. Although performance attributes are always with crisp values, anomaly attributes cover both crisp and fuzzy values owing to the ambiguity or uncertainty during information capturing stage. Variables with crisp values are the attributes that are calculated according to a traditional set, that is, the membership of any value for the attributes is either 1 or 0. Variable with a fuzzy value defines their value with the membership that varied from 0 to 1, which are always determined by the membership function.

Here, the triangular membership function is used to convert the numerical data into fuzzy values. For example, if feature A is valued as numeral x, the values of feature A for all items $u \in U$ can be denoted as $X = \{x(u), u \in U\}$. Then,

X can be grouped to k semantic clusters T_i, $i \in \{1, \ldots, k\}$. Semantic clusters T_i has a triangular membership function as:

$$u_{T_1(x)}(x) = \begin{cases} 1, x \le m_1 \\ (m_2 - x)/(m_2 - m_1), m_1 < x < m_2 \\ 0, x \ge m_2 \end{cases} \tag{7.3}$$

$$u_{T_k(x)}(x) = \begin{cases} 1, x \ge m_k \\ (x - m_{k-1})/(m_k - m_{k-1}), m_{k-1} < x < m_k \\ 0, x \le m_{k-1} \end{cases} \tag{7.4}$$

$$u_{T_i(x)}(x) = \begin{cases} 0, x \ge m_{i+1} \\ (m_{i+1} - x)/(m_{i+1} - m_i), m_i < x < m_{i+1}, 1 < i < k \\ (x - m_{i-1})/(m_i - m_{i-1}), m_{i-1} < x < m_i \\ 0, x \le m_{i-1} \end{cases} \tag{7.5}$$

where, m_i, $i \in \{1, \ldots, k\}$ denotes the ith center of semantic clusters.

7.6.4 Tree Builder

Mainly three steps of rule induction are needed in DT learning. Step 1: create a large DT from historical cases according to feature selection approaches; Step 2: prune the branches and nodes that have little statistical influence on the tree; and Step 3: improve the understandability of obtained tree. Since the DT technologies have been widely discussed, we give the feature selection approaches briefly.

Fuzzy-ID3 is an extension of the classical tree-building ID3 algorithm and it has the ability to analyze both crisp and fuzzy variables. Fuzzy-ID3 algorithm is similar to ID3; the difference is that while ID3 selects the test attribute based on the information gain computed by using the probability of ordinary data, Fuzzy-ID3 does that by using the likelihood of membership values for the data set.

Assume a set of data E, and each data has r kinds of attributes and one classified class $S = \{S_1, \ldots, S_n\}$ and fuzzy sets $F = \{F_{i1}, F_{i2}, \ldots, F_{im}\}$ for each attribute A_i, m is the number of clusters for A_i. Let E^{S_n} be a fuzzy/crisp set in E which has S_n kinds of class and $|E|$ be the total number of membership or crisp values of the set of data E. Then, the information gain $G(A,E)$ for attribute A by a fuzzy/crisp set of data E is described as:

$$G(A, E) = I(D) - \text{Entropy}(A, E) \tag{7.6}$$

where

$$I(D) = -\sum_{i=1}^{n} (p_k \bullet \log_2 p_k) \tag{7.7}$$

$$\text{Entropy}(A,E)=\sum_{j=1}^{m}(p_{ij}\cdot I(E_{F_{ij}})) \tag{7.8}$$

$$p_k=\frac{|E^{S_n}|}{|E|} \tag{7.9}$$

$$p_{ij}=\frac{|E_{F_{ij}}|}{\sum_{j=1}^{m}|E_{F_{ij}}|} \tag{7.10}$$

For a case where identical information gains are calculated, one is either selected randomly or chosen based on its importance within the project, and the DT is generated step by step until all the nodes are chosen.

C4.5 can only handle the crisp set and it applies the "information gain ratio" to obtain splits at the intermediate nodes in the tree:

$$\text{Gain ratio}=\frac{\text{Gain}(A,E)}{\text{Split}(E)} \tag{7.11}$$

where,

$$\text{Split}(E)=-\sum_{i=1}^{S}\frac{|E_i|}{|E|}\times\log_2\left(p\frac{|E_i|}{|E|}\right) \tag{7.12}$$

If a is an attribute that carries continuous value, the attribute need to be discretized into two intervals afore the split.

7.6.5 Anomaly Extraction and Causes Diagnosis

After a real-time PP case is acquired, the rules correlated to the information can be used to assess the PP condition. If one anomaly event is detected, the anomaly label (e.g., Label L1) will be affixed to it, thus the following anomaly cause diagnosis process can easily identify the anomaly.

To detect causes for a new anomaly case, this chapter uses the reverse maximum matching (RMM) approach. After a new anomaly event is found, the information correlated to the anomaly, that is, all the real-time manufacturing resources data and fuzzy rules, are involved from the database. Then, the rules which have the same anomaly degree with the new anomaly are retained. Finally, based on the membership and certainty of historical rules, the causes are obtained. The main processes for the anomaly diagnosis are presented as follows:

1. Compute matching membership between the real-time PP case and each decision rule, and the result is deemed as the fitting membership between the case and the rule.
2. If only one rule have the highest fitting membership, this rule is chosen as the cause.

3. If several rules share the same highest fitting membership, the rule with the highest certainty is chosen.

Finally, the production anomaly analysis result is uploaded to the evaluation managers for acceptability assessment. If the anomaly analysis case is acknowledged, the anomaly can be dealt according to the causes, and the new case can be added into the database. Through the update of the anomaly cases and new knowledge supplemented, modifications on the database are necessary to ensure the knowledge is up to date and reliable. As a result, good base for future optimal decision making is laid.

REFERENCES

[1] Y. Zhang, G. Zhang, J. Wang, S. Sun, S. Si, T. Yang, Real-time information capturing and integration framework of the Internet of manufacturing things, Int. J. Comput. Integr. Manuf. 3052 (December) (2014) 1–12.

[2] L. Da Xu, W. He, S. Li, Internet of things in industries: a survey, IEEE Trans. Ind. Inform. 10 (4) (2014) 2233–2243.

[3] T. Sánchez López, D.C. Ranasinghe, M. Harrison, D. McFarlane, Adding sense to the Internet of things: an architecture framework for smart object systems, Pers. Ubiquit. Comput. 16 (3) (2012) 291–308.

[4] F. Tao, Y. Cheng, L. Da Xu, L. Zhang, B.H. Li, CCIoT-CMfg: cloud computing and Internet of things-based cloud manufacturing service system, IEEE Trans. Ind. Inform. 10 (2) (2014) 1435–1442.

[5] T. Murata, Petri nets: properties, analysis and applications, Proc. IEEE 77 (4) (1989) 541–580.

[6] X. He, T. Murata, High-level Petri nets—extensions, analysis, and applications", The Electrical Engineering Handbook, Elsevier Academic Press, Amsterdam, 2005, pp. 459–475.

[7] R. Kohavi, R. Quinlan, Decision Tree Discovery, Handb. Data Min. Knowl. Discov. 3 (Hunt 1962) (1999) 267–276.

[8] R.C. Barros, M.P. Basgalupp, A.C.P.L.F. De Carvalho, A.A. Freitas, A survey of evolutionary algorithms for decision-tree induction, IEEE Trans. Syst. Man Cybernet. C 42 (3) (2012) 291–312.

[9] Z.X. Guo, E.W.T. Ngai, C. Yang, X. Liang, An RFID-based intelligent decision support system architecture for production monitoring and scheduling in a distributed manufacturing environment, Int. J. Prod. Econ. 159 (2015) 16–28.

[10] Y. Chai, Y. Zhou, Y Wang. Collaborative manufacturing execution systems: Using real-time information to support lean production, Computer Supported Cooperative Work in Design, 2008. CSCWD 2008. 12th International Conference on. IEEE, 2008, pp. 857–861.

[11] X.V. Wang, X.W. Xu, An interoperable solution for cloud manufacturing, Robot. Comput. Integr. Manuf. 29 (4) (2013) 232–247.

[12] L. Wang, P. Orban, A. Cunningham, S. Lang, Remote real-time CNC machining for web-based manufacturing, Robot. Comput. Integr. Manuf. 20 (6 SPEC) (2004) 563–571.

[13] R.-S. Chen, M.A. Tu, J.-S. Jwo, An RFID-based enterprise application integration framework for real-time management of dynamic manufacturing processes, Int. J. Adv. Manuf. Technol. 50 (9–12) (2010) 1217–1234.

[14] J. Chongwatpol, R. Sharda, RFID-enabled track and traceability in job-shop scheduling environment, Eur. J. Oper. Res. 227 (3) (2013) 453–463.

[15] I. Arkan, H. Van Landeghem, Evaluating the performance of a discrete manufacturing process using RFID: a case study, Robot. Comput. Integr. Manuf. 29 (6) (2013) 502–512.
[16] M. Bevilacqua, F.E. Ciarapica, A. Crosta, G. Mazzuto, C. Paciarotti, Implementation of a RFID system in a furniture industry involved in the fashion sector: a case study, Int. J. RF Technol. Res. Appl. 6 (2–3) (2015) 99–119.
[17] P. Georgiadis, C. Michaloudis, Real-time production planning and control system for job-shop manufacturing: a system dynamics analysis, Eur. J. Oper. Res. 216 (1) (2012) 94–104.
[18] C. Zang, Y. Fan, Complex event processing in enterprise information systems based on RFID, Enterp. Inf. Syst. 1 (1) (2007) 3–23.
[19] G.Q. Huang, Y. Zhang, Q.Y. Dai, O. Ho, F.J. Xu, Agent-based workflow management for RFID-enabled real-time reconfigurable manufacturing, Collaborative Design and Planning for Digital Manufacturing, Springer, London, 2009, pp. 341–364.
[20] Y. Zhang, T. Qu, O. Ho, G.Q. Huang, Real-time work-in-progress management for smart object-enabled ubiquitous shop-floor environment, Int. J. Comput. Integr. Manuf. 24 (5) (2011) 431–445.
[21] S. Kumaraguru, B. (Serm) Kulvatunyou, K.C. Morris, Integrating real-time analytics and continuous performance management in smart manufacturing systems, IFIP Adv. Inf. Commun. Technol. 440 (Part 3) (2014) 175–182.
[22] L.M. Blaha, C.R. Fisher, M.M. Walsh, B.Z. Veksler, G. Gunzelmann. Real-time fatigue monitoring with computational cognitive models. International Conference on Augmented Cognition, Springer International Publishing, Toronto, 2016, pp. 299–310.
[23] J. Li, Performance analysis of production systems with rework loops, IIE Trans. 36 (2004) 1–41.
[24] M. Colledani, T. Tolio, Integrated analysis of quality and production logistics performance in manufacturing lines, Int. J. Prod. Res. 49 (2) (2011) 485–518.
[25] Y. Lao, L. Liu, Performance evaluation of bus lines with data envelopment analysis and geographic information systems, Comput. Environ. Urban Syst. 33 (4) (2009) 247–255.
[26] Y.-K. Lin, P.-C. Chang, J.C. Chen, Performance evaluation for a footwear manufacturing system with multiple production lines and different station failure rates, Int. J. Prod. Res. 51 (January) (2012) 1–15.
[27] R.Y. Zhong, G.Q. Huang, Q.Y. Dai, T. Zhang, Mining SOTs and dispatching rules from RFID-enabled real-time shopfloor production data, J. Intell. Manuf. 25 (4) (2014) 825–843.
[28] Y. Lv, C.K.M. Lee, H.K. Chan, W.H. Ip, RFID-based colored Petri net applied for quality monitoring in manufacturing system, Int. J. Adv. Manuf. Technol. 60 (1–4) (2012) 225–236.
[29] N. Wu, Z. Mengchu, Intelligent token Petri nets for modelling and control of reconfigurable automated manufacturing systems with dynamical changes, Trans. Inst. Meas. Control 33 (1) (2011) 9–29.
[30] Y. Zhang, W. Wang, N. Wu, C. Qian, IoT-enabled real-time production performance analysis and exception diagnosis model, IEEE Trans. Autom. Sci. Eng. 13 (2015) 1–15.
[31] Ö. Başak, Y.E. Albayrak, Petri net based decision system modeling in real-time scheduling and control of flexible automotive manufacturing systems, Comput. Ind. Eng. 86 (2014) 116–126.
[32] C. Wang, X. Feng, X. Li, X. Zhou, P. Chen, Colored Petri net model with automatic parallelization on real-time multicore architectures, J. Syst. Archit. 60 (3) (2014) 293–304.
[33] X. Meng, Modeling of reconfigurable manufacturing systems based on colored timed object-oriented Petri nets, J. Manuf. Syst. 29 (2–3) (2010) 81–90.
[34] S. Ha, H.-W. Suh, A timed colored Petri nets modeling for dynamic workflow in product development process, Comput. Ind. 59 (2–3) (2008) 193–209.
[35] L. Evans, N. Lohse, M. Summers, A fuzzy-decision-tree approach for manufacturing technology selection exploiting experience-based information, Expert Syst. Appl. 40 (16) (2013) 6412–6426.

[36] J. Zarei, M.A. Tajeddini, H.R. Karimi, Vibration analysis for bearing fault detection and classification using an intelligent filter, Mechatronics 24 (2) (2014) 151–157.
[37] F.G. Cabral, M.V. Moreira, O. Diene, J.C. Basilio, A Petri net diagnoser for discrete event systems modeled by finite state automata, IEEE Trans. Automat. Contr. 60 (1) (2015) 59–71.
[38] M.P. Cabasino, A. Giua, M. Pocci, C. Seatzu, Discrete event diagnosis using labeled Petri nets. An application to manufacturing systems, Control Eng. Pract. 19 (9) (2011) 989–1001.
[39] S.-G. He, Z. He, G.A. Wang, Online monitoring and fault identification of mean shifts in bivariate processes using decision tree learning techniques, J. Intell. Manuf. 24 (1) (2013) 25–34.
[40] A. Maier, A. Vodencarevic, O. Niggemann, R. Just, M. Jaeger, Anomaly detection in production plants using timed automata, in: Eighth International Conference on Informatics in Control, Automation and Robotics (ICINCO), 2011, pp. 363–369.
[41] S. Faltinski, H. Flatt, F. Pethig, B. Kroll, A. Vodenčarević, A. Maier, O. Niggemann, Detecting anomalous energy consumptions in distributed manufacturing systems, in: IEEE International Conference on Industrial Informatics (INDIN), 2012, pp. 358–363.
[42] S. Yin, H. Luo, S.X. Ding, Real-time implementation of fault-tolerant control systems with performance optimization, IEEE Trans. Ind. Electron. 61 (5) (2014) 2402–2411.
[43] M. Chioua, M. Bauer, S.-L. Chen, J.C. Schlake, G. Sand, W. Schmidt, N.F. Thornhill, Plant-wide root cause identification using plant key performance indicators (KPIs) with application to a paper machine, Control Eng. Pract. 49 (2016) 149–158.
[44] T. Wang, G. Zhang, J. Zhao, Z. He, J. Wang, M.J. Perez-Jimenez, Fault diagnosis of electric power systems based on fuzzy reasoning spiking neural p systems, IEEE Trans. Power Syst. 30 (3) (2015) 1182–1194.
[45] J.W.D. Groenewald, C. Aldrich, Root cause analysis of process fault conditions on an industrial concentrator circuit by use of causality maps and extreme learning machines, Miner. Eng. 74 (2015) 30–40.
[46] J. Hu, L. Zhang, Z. Cai, Y. Wang, A. Wang, Fault propagation behavior study and root cause reasoning with dynamic Bayesian network based framework, Process Saf. Environ. Prot. 97 (2015) 25–36.
[47] A. Alaeddini, I. Dogan, Using Bayesian networks for root cause analysis in statistical process control, Expert Syst. Appl. 38 (9) (2011) 11230–11243.
[48] S. Yin, X. Zhu, O. Kaynak, Improved PLS focused on key-performance-indicator-related fault diagnosis, IEEE Trans. Ind. Electron. 62 (3) (2015) 1651–1658.
[49] G. Li, T. Yuan, S.J. Qin, T. Chai, Dynamic time warping based causality analysis for root-cause diagnosis of nonstationary fault processes, IFAC—PapersOnLine 48 (8) (2015) 1288–1293.
[50] C.Z. Janikow, Fuzzy decision trees: issues and methods, IEEE Trans. Syst. Man. Cybern. B 28 (1) (1998) 1–14.

Chapter 8

Real-Time Information-Driven Production Scheduling System

8.1 INTRODUCTION

Production scheduling refers to the process of allocating right manufacturing resources to specific time periods to complete a set of manufacturing processes in the plan [1]. As an important manufacturing planning activity, it aims to tackle resource utilization and time span of the manufacturing operations. Agent-based manufacturing scheduling systems [2] provide a promising way for production scheduling. With the rapid developments of information communication technologies (e.g., wireless sensors, wireless sensor network), ubiquitous manufacturing (UM) has been proposed as one core manufacturing technology in advanced manufacturing systems [3,4]. Ubiquitious manufacturing makes the real-time status of manufacturing resources can be easily captured, then promote real-time production scheduling. Meanwhile, the unscheduled downtime in a UM system can reach a lowest level by using the captured real-time information, which provides one more effective way to deal with negative impact of production exceptions [5].

Though significant progress has been made in the field of real-time information capturing and processing, some research questions still need to be considered in how to implement real-time scheduling methods in real-life manufacturing floors.

1. During the process planning stage, real-time statuses of manufacturing resources are rarely considered when tasks are allocated to machines. The mentioned circumstance may lead to nonoptimal allocation of tasks and have negative influence on following production scheduling.
2. The scarcity of effective methods to capture and process real-time manufacturing information leads to the inefficient, inaccurate, and time-delay production monitor. Meanwhile, an effective mechanism has not been well developed to accomplish the real-time information integration between production scheduling stage and production execution exception monitor stage. This imposes more challenges on performing real-time production scheduling.

Optimization of Manufacturing Systems Using the Internet of Things
http://dx.doi.org/10.1016/B978-0-12-809910-0.00008-1

3. Nowadays, tasks in the manufacturing system show the features of large variety and small batch size. This phenomenon raises more requirements in the field of real-time production monitoring. It has to own the capabilities of dynamic decision making and adaptive control to tackle different production changes. Thus it is very important to develop real-time production scheduling architecture and method for the dynamic changing manufacturing environment.

To address previously mentioned questions, in this section, the promising advantages of advanced information communication technologies and multiagent are integrated to perform real-time production scheduling in a UM environment. The developed framework of real-time information-driven production scheduling (RIDPS) system aims to bridge the gap between production planning and execution during the whole manufacturing process. This chapter also aims to provide a way for enhancing the productivity and flexibility of manufacturing system through implementing the proposed real-time information-driven production system.

The rest of this chapter is organized as follows. Section 8.2 reviews related works in three different sides. The framework of RIDPS system is designed in Section 8.3. Multimodels, such as equipment agent (EA), capability evaluation agent (CEA), real-time scheduling agent (RSA), and production execution monitor agent (PEMA) are described in Sections 8.4–8.7, respectively. Section 8.8 explains the genetic algorithm (GA)-based production scheduling algorithm.

8.2 RELATED WORKS

Three streams of works are relevant to this research. These include (1) agent technology and applications in manufacturing field, (2) real-time production scheduling, and (3) manufacturing information monitor technology.

8.2.1 Agent Technology and Applications in Manufacturing Field

Agent technology is an important subfield of artificial intelligence (AI), and it is one of the powerful technologies for the development of large-scale distributed systems to deal with the uncertainty in a dynamic environment [6–8]. A new inter- and intraagent cooperation approach was presented to improve the performance of multiagent or distributed manufacturing systems [7]. Sikora and Shaw described a model for the coordination and integration of enterprise information systems (EISs), which can model typical EISs as containing multiple agents with different functionalities [9]. An agent-based system for coordinated product development and manufacture was presented by Jia et al.; the system consists of two classes of agents, that is, managing agent as the core manager and functional agents with specific functionality [10]. An agent-based multicontract negotiation system was proposed to address the challenges of global manufacturing supply chain coordination [11]. Zhang et al. presented a novel gateway technology for the real-time management in UM environment; the agent-based smart objects management system was used to manage the multiple auto-ID sensors to capture the real-time status [4,12]. Farid and Ribeiro presented a multiagent system reference architecture for reconfigurable industrial systems based on a quantitative and formal design method, which was embedded in an traditional engineering design methodology called axiomatic design for large flexible engineering systems [13]. Trappey et al. proposed an agent-based co-operative mold production system, which aimed to sustain the collaborative and self-directed mold manufacturing outsourcing patterns [14]. Kumari et al. presented an autonomous self-adaptive multiagent system to aid small and medium enterprises to obtain the optimal decision so that the uncertainty in supply chain can be alleviated [15]. Wang et al. proposed a distributed multiagent reinforcement learning algorithm to solve the resource-constrained imperfect preventive maintenance question; the question was modeled as a semi-Markov decision process [16]. Vogel-Heuser et al. proposed a method that used the rapidly developing concept of cyber-physical systems for a case of manufacturing systems by methods of software agents [17]. Manupati et al. developed a multiobjective mathematical model in the background of networked manufacturing systems, and a mobile agent–based negotiation pattern was presented for solving the distributed problem [18]. Vidoni and Vecchietti designed a smart agent for ERP's data structure analysis based on ANSI/ISA-95 standard; three open source ERPs: OpenERP, Dolibarr, and Adempiere were considered as the cases [19]. Ayhan et al. proposed a multiagent-based approach for changing management mode in industrial companies [20]. Bearzotti et al. designed a self-governing multiagent method to manage the supply chain event, which can perform autonomous corrective control activities to minimize the influence of deviations of the currently executing plan [21].

8.2.2 Real-Time Production Scheduling

The scheduling problem in shop floor represents a problem where the objective is to properly allocate available resources to tasks in order to optimize an objective function, which is usually related to time, like the makespan [22], total completion time [23], tardiness [24], maximum lateness, total throughput time [25], etc. Previous approaches mainly focus on the production scheduling before the production starts or only research under a theoretical environment. For example, Aghezzaf presented the production planning and warehouse management for supply networks with interfacility mold transfers [26]. Babayan and He presented an overall methodology of agent-based manufacturing systems scheduling, incorporating game theoretic analysis of agent cooperation, to solve the *n*-job 3-stage flexible flow shop scheduling problem [27]. Fang et al. provided a new mixed-integer linear programming model for scheduling a typical flow shop that combined the peak total power consumption and the associated carbon footprint with the makespan [28]. Luh and Chueh presented an innovative multimodal immune algorithm for discovering optimal solutions to job shop scheduling problems imitating the features of a biological immune system [29]. Giordani et al. presented a distributed multiagent production planning and scheduling framework for mobile robots [30]. Fan et al. considered the problem of integrated scheduling of production and delivery on a single machine, in which the availability constraint of the machine and jobs in processing may be disturbed [31]. Shishvan and Sattarvand discussed the long-term production planning problem of open pit mines by ant colony optimization method [32]. Cheng et al. considered an integrated scheduling of production and distribution so as to minimize the production and distribution costs, and an improved ant colony optimization method is used to solve the problem [33]. Arauzo et al. investigated a new method based on multiagent systems and a combinatorial auction mechanism was used to allocate resources for the projects tasks [34].

As to the real-time scheduling aspect, Anderson and Calandrino proposed a scheduling method for real-time systems, which is realized on multicore platforms, so that individual threads of multithreaded real-time tasks can be scheduled together [35]. Tabuada presented an event-triggered real-time scheduling method for stabilizing control tasks, where the real-time scheduler were regarded as a feedback controller that decided which task was performed at any given time [36]. Subramanian et al. analyzed real-time scheduling algorithms for coordinated aggregation of deferrable loads and storage. Three scheduling policies: earliest deadline first, least laxity first, and receding horizon control were discussed in the paper, and the performance of those algorithms were studied through simulations [37]. Buyurgan and Saygin proposed a multicriteria decision-making framework for real-time scheduling and part routing solutions by implementing pairwise comparison of possible future states of a manufacturing system [38]. Yan and Wang proposed a two-layer dynamic scheduling approach for the dynamic scheduling problem of a reentrant production line, in which all of the parts are assumed to have the same processing routes and need to be processed on every machine [39]. Du et al. presented a new framework for integrating, scheduling, and nonlinear control for continuous processes, and the approach for reducing dimensions of the problem and closed-loop process dynamics were considered [40]. Lee and Prabhu proposed a dynamic algorithm for distributed feedback control, which considered the functions of production and maintenance scheduling at the shop-floor level and machinery capacity control at the CNC level at the same time, while the two problems were usually considered in isolation in practice [41]. Luo et al. discussed the real-time scheduling problem for hybrid flow shop in UM environment; a multiperiod hierarchical scheduling method is presented in the paper [42]. Based on a DT model, Choi et al. presented a real-time scheduling method for reentrant hybrid flow shops [43]. Unlike the theoretical approach on reentrant hybrid flow shop scheduling, a real-time scheduling approach using a decision tree when selecting appropriate dispatching rules was creatively provided.

8.2.3 Manufacturing Information Monitor Technology

Thanks to the emerging advanced IoT technologies, more and more manufacturing enterprises began a widespread use of IoT technologies to implement and manage their business [44,45]. The IoT provides an IT infrastructure to facilitate the information exchange of "things and process" in a real-time and reliable way. Some scholars have explored IoT technologies practice in manufacturing information monitor.

Huang et al. proposed a theoretical wireless manufacturing (WM) framework [46]; WM relied substantively on wireless devices, such as radio frequency identification (RFID) or auto-ID sensors and wireless information networks for real-time monitoring of production data. Zhang et al. described a real-time information capturing and integration architecture of the Internet of manufacturing things [44], and a real-time manufacturing information integration service was proposed to fulfill the target of timely information exchange among EISs and manufacturing resources in shop-floor frontline. Based on RFID technology, Wang et al. proposed a digital warehouse management system in the tobacco industry [47]. By using RFID technology, the system enabled a plane warehouse to achieve visualized inventory management, automatic storage assignment, and high accuracy of inventory control. Combining RFID technology with ontologies, Grüninger et al. created smart objects in the context of manufacturing process to solve the problems of massive RFID tags interoperability [48]. Xu et al. focused on the research of closed-loop product information tracking and feedback in a wireless technology–enabled environment from the modeling point of view [49]. By using 2D barcode and RFID technologies, Lin et al. proposed a novel system called Mobile 2D Barcode/RFID-based Maintenance Management system to improve lab equipment and instrument maintenance management and provides a maintenance information sharing platform [50]. Subramaniam et al. presented an automated data collection and display structure for production lines; the system can generate an automated report, which stayed in place and the management only needed to act based on the results [51].

8.3 OVERALL ARCHITECTURE OF REAL-TIME INFORMATION-DRIVEN PRODUCTION SCHEDULING SYSTEM

The overall architecture of RIDPS is shown in Fig. 8.1. The objective of RIDPS is to achieve the dynamic optimization of process tasks based on the real-time status of the equipment. By extending the IoT technologies into the traditional manufacturing field, the status of manufacturing resources can be sensed timely. Then, the tasks can be allocated to the equipment based on the real-time capability of machines. When production anomaly happens, such as machine broken, order canceling, and so on, the rescheduling can be executed according to the real-time production information.

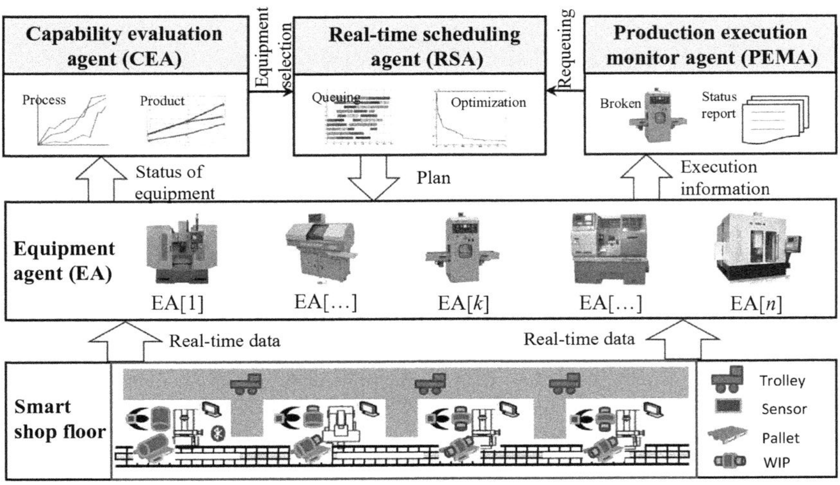

FIGURE 8.1 Overall architecture of real-time information-driven production scheduling (RIDPS).

Four kinds of agents are designed in this chapter to achieve real-time production scheduling. They are briefly described as follows:

1. EA
   It is used to actively acquire the real-time status of the equipment. This agent is the foundation of the real-time production scheduling. Some auto-ID devices are used to sense the raw production data, and then the captured data are integrated to reflect the meaningful equipment status.
2. CEA
   It is responsible for evaluating the capability of the equipment according to the real-time status sensed by machine agent. Then, the process planning can choose an optimal equipment for each process of the tasks.
3. RSA
   This agent is used to make an optimal plan, which can allocate tasks to the available equipment with the optimized objective function. First, the mathematic model is given; the objective function and constraints are designed based on the real-time production problem. Then, the intelligent algorithm is given to solve the scheduling or rescheduling problem.
4. PEMA
   It aims to sensing the real-time data of the various kinds of manufacturing resources. During shop-floor frontline, production anomaly often happens; it is necessary to track and trace the anomaly in order to achieve the target for real-time production scheduling.

8.4 EQUIPMENT AGENT

EA is used to wrap the applications of equipment side and process the multi-source real-time data captured from wireless devices (e.g., RFID). The functions of EA can be categorized into two aspects: (1) it can centrally manage the different kinds of auto-ID devices, which are deployed on the equipment side. These auto-ID devices can capture real-time manufacturing data based on a specific working logic; (2) EA can transform the captured manufacturing data into useful manufacturing information and provide the manufacturing information for other modules in the real-time production scheduling system.

Fig. 8.2 demonstrates the EA model. It consists of two components, namely data capturing and application service.

1. Data capturing
   Auto-ID devices are deployed on the machine side. They are used to capture dynamic manufacturing data. Data capturing component aims to manage these auto-ID devices. It includes two modules.
   The first module is named Definition and is auto driven. It is used to store various drivers of heterogeneous auto-ID devices and form a driver pool. This driver pool makes new deployed auto-ID devices present a mode of "plug and play."
   The second module is named standard data capturing module. This module aims to unify the data perception functions of different auto-ID devices so

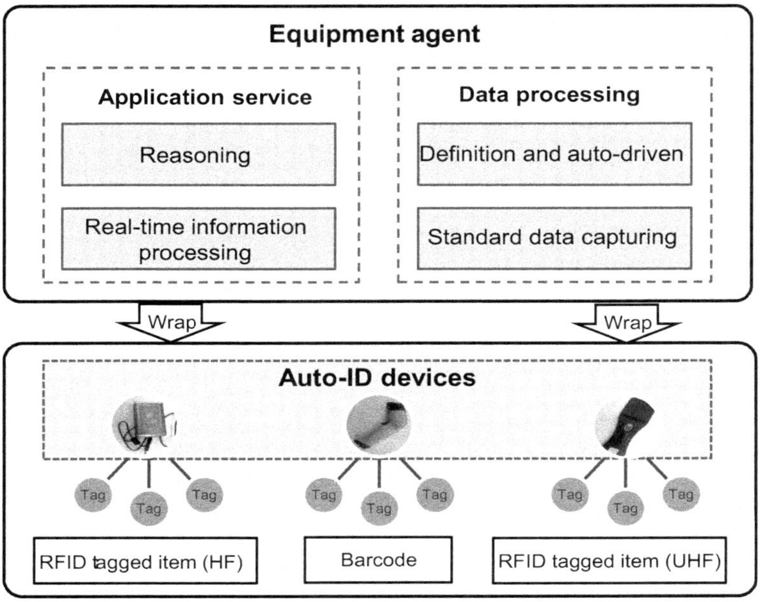

FIGURE 8.2 Machine agent model.

that the capturing data can be easily invoked with a standard way. For example, the perception function can be stated as follows, "readingData (Parameter [1], Parameter [i])" and "writingData (Parameter [1], Parameter [i])."

2. Application services
   This component is responsible for providing value-added manufacturing information for other modules in the real-time production scheduling system. It also includes two modules.
   The first module is named reasoning module. It is used to improve the intelligence of EA. By this module, EA can learn which kind of resource is coming or leaving the equipment. In order to promote the decision making of EA, rule-based methods are installed in this module.
   The second module is named real-time information processing. This module aims to process various captured real-time data. Comparing with the reasoning module, this module mainly focuses on getting useful real-time manufacturing information. For example, the "getMaterial()" will return detailed real-time information, such as the kinds of material, real-time quantity of required material, and so on.

8.5 CAPABILITY EVALUATION AGENT MODEL

Capability evaluation agent (CEA) is responsible for optimally allocating processes of tasks to related manufacturing equipment according to real-time manufacturing information provided by EA. The model of CEA is demonstrated in Fig. 8.3. Assume that a manufacturing task includes "*n*" processes, and a group of machines in the shop floor can meet the manufacturing requirements of each process [i]. CEA

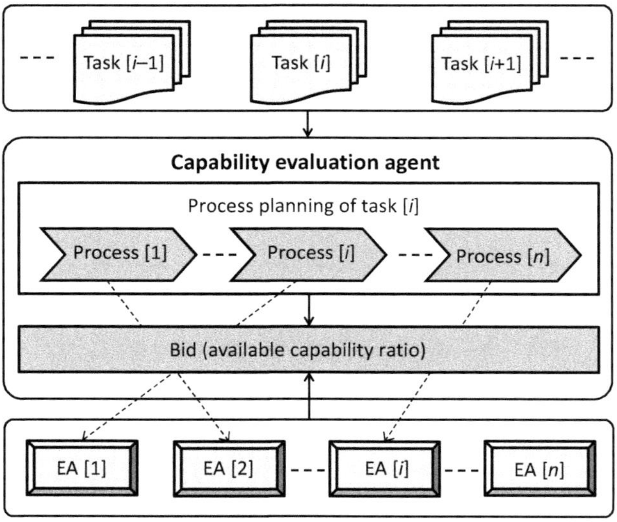

FIGURE 8.3 Capability evaluation agent (CEA) model.

aims to find an optimal machine from the candidate machines to complete process [i] through a bid mechanism. Traditional allocation methods rarely consider real-time workloads of machines and the changing manufacturing environment. The manufacturing may be invalid or nonoptimal during the execution process.

In the model of CEA, capability ratio is employed to evaluate the capabilities of machines and find the optimal one for the process [i]. As said before, a bid mechanism is used to accomplish the selection of optimal machine. First, a group of candidate machines defined as g is formed. Then, the related EAs of candidate machines will bid to complete process [i] based on their real-time capabilities. Finally, the available capabilities of candidate machines will be calculated by CEA, and the optimal manufacturing machine will be selected according to objective function Eq. (8.1).

$$\mathrm{Max}\{ACR_m \mid m \in g\} \tag{8.1}$$

where, $ACR_m = AC_m / TC_m \times 100\%$.

Here, ACR_m is defined as the available capability ratio of machine "m". AC_m represents the real-time available capability of machine "m"; it is changing dynamically with allocated tasks queue. TC_m indicates the total capability of machine "m"; it is constant.

8.6 REAL-TIME SCHEDULING AGENT MODEL

Fig. 8.4 presents the overview of RSA; it assigns the processes of tasks to the equipment based on the real-time manufacturing information from CEA and PMA. After the scheduling, the EA can obtain the task queue, which can be

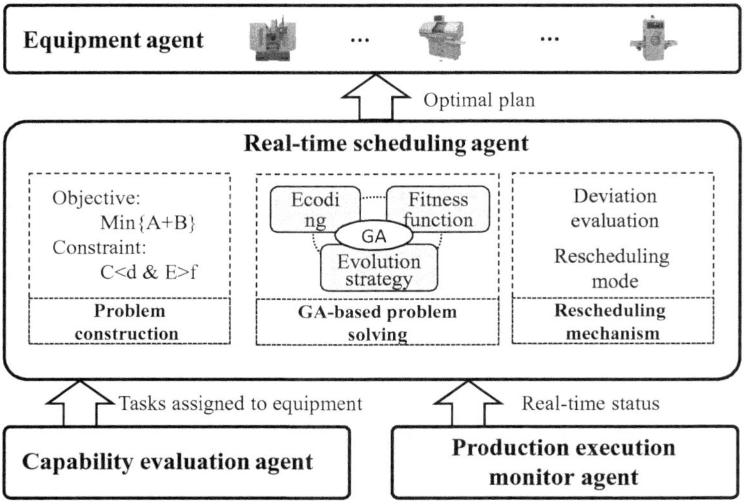

FIGURE 8.4 Real-time scheduling agent (RSA) model.

represented as {*i*, *j*, *k*, *BT*, *CT*}. Here, {*i*, *j*, *k*, *BT*, *CT*} denotes the equipment "*k*" which is allocated to finish the *j*th process of task "*i*" with the begin time "*BT*" and completed time "*CT*."

Three modules, namely problem construction modules, solving module, and rescheduling module are designed in RSA.

1. Problem construction
   Before the mathematic formulation are provided, the corresponding notations are given at first. The details can be shown in Table 8.1.
   According to the notations, the mathematic model is constructed as follows.
   Objective function:

$$\mathrm{Min}\left\{\sum_{i=1}^{n}\mu_i\cdot\max(0, C_i - D_i)\right\} \tag{8.2}$$

Subject to:

$$BT(T_i, TP_{j+1}, E_a) - BT(T_i, TP_j, E_b) \geq PT(T_i, TP_j, E_b) \tag{8.3}$$

$$\begin{array}{l} BT(T_x, TP_c, E_k) - BT(T_y, TP_d, E_k) \geq PT(T_y, TP_d, E_k) \text{ or} \\ BT(T_y, TP_d, E_k) - BT(T_x, TP_c, E_k) \geq PT(T_x, TP_c, E_k) \end{array} \tag{8.4}$$

$$i, x, y \in [1, n]\, j \in [1, N_i], c \in [1, N_x], k, a, b \in [1, m]$$

Eq. (8.2) represents the objective function, which takes the maximal total tardiness cost of all the tasks into consideration. Eq. (8.3) denotes processes

TABLE 8.1 Notations

Notations	Description
$E = \{E_1, \ldots, E_k, \ldots, E_m\}$	Set of equipment, *m* denotes the total number of equipment
$T = \{T_1, \ldots, T_i, \ldots, T_n\}$	Set of tasks, *n* represents the total number of tasks
$TP_i = \{tp_1, tp_2, \ldots, tp_{N_i}\}$	Set of processes of task "*i*," N_i denotes the total number of processes of task "*i*"
(T_i, TP_j, E_k)	The process "*j*" of task "*i*" will be processed at equipment "*k*"
$BT(T_i, P_j, E_k)$	Begin time of (Ti, TP_j, E_k)
$PT(T_i, P_j, E_k)$	Process time of (T_i, TP_j, E_k)
$C = \{C_1, \ldots, C_i, \ldots, C_n\}$	Set of complete time of task "*i*"
$D = \{D_1, \ldots, D_i, \ldots, D_n\}$	Set of due date of task "*i*"

order constraint, which means that the later process of task need to begin after the prior one. Eq. (8.4) denotes resource constraint, that is, the process of different tasks cannot be processed at the same equipment at the same time.

2. GA-based solving module
This module aims to find an optimal production queue based on established mathematic problem by using the intelligent algorithm. Since GAs are adaptive methods that are widely studied and used for solving optimization problems in manufacturing fields, they are adopted to solve the aforementioned problem. More details of the GA-based solving method are described in Section 8.8.
3. Rescheduling module
As discussed before, production anomaly is unavoidable, rescheduling is essential in maximizing the production capability. During the execution stage, the PMA will feedback the production information on time. The total deviation ratio of the machines at time (t) will be computed as shown in Eq. (8.5)

$$\Delta(t)=\sum_{k=1}^{m}\frac{w1\times AT^{t}(i,j)+w2\times LT^{t}(i,j)+ET^{t}(k)}{DT^{t}(k)}\times 100\% \tag{8.5}$$

where, $w1$ and $w2$ are weights, generally, $w1$ is less than 1 and $w2$ is greater than 1; $AT^t(j)$ is the ahead time of the jth process of task (i) assigned to equipment (k) at time (t) contrast to the previous plan; $LT^t(j)$ is the tardiness time of the jth process of task (i) assigned to equipment (k) at time (t) contrast to the previous plan; $ET^t(k)$ is the anomaly handling time of the equipment (k); $DT^t(k)$ is the duration time of equipment (k). In the previous plan, let T be the completed time of the last process assigned to equipment (k), then $DT^t(k)$ can be obtained by subtracting current time (t) from T.
According to the calculation of $\Delta(t)$, the rescheduling approach can be executed. Two manners, namely local rescheduling and global rescheduling, are presented in this module. On one hand, if $\Delta(t) \leq \lambda$, the local rescheduling manner is triggered. It is applied to requeuing a small quantity of the tasks while other tasks' queue is not altered. It deals with the anomalies, such as the queue anomaly of an equipment. On the other hand, if $\Delta(t) > \lambda$, the global rescheduling manner is triggered. It is applied to requeuing all the tasks from the anomaly time (t) by using the procedure designed in GA-based solving module. Here, λ is a constant and is default set as 10%, and it can vary based on different managers.

8.7 PRODUCTION EXECUTION MONITOR AGENT MODEL

Fig. 8.5 shows the model of PEMA. It is responsible for analyzing the real-time information from EAs and sending real-time manufacturing information to RSA.

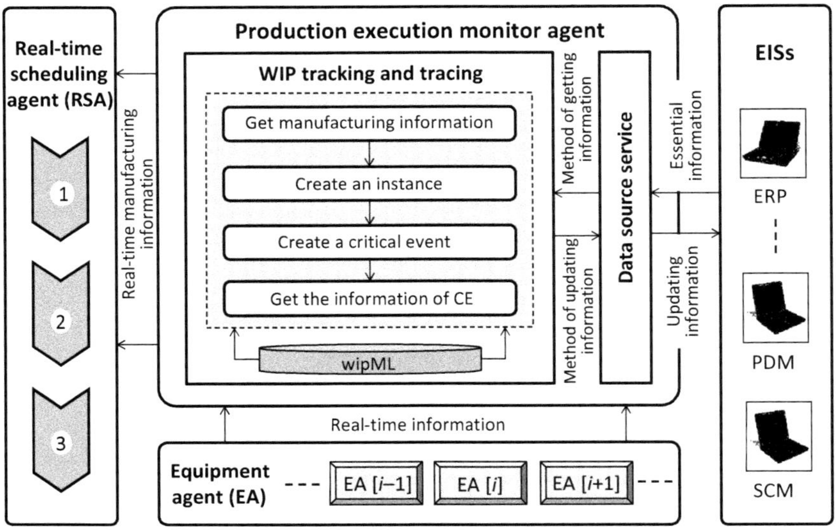

FIGURE 8.5 Production execution monitor agent (PEMA) model.

The working logic of PEMA is composed of three stages. At the first stage, data source service is invoked to obtain essential information [e.g., bill of material (BOM)] about related production task. The essential information is stored in the up-level enterprise information management systems. At the second stage, a new work in progress (WIP) instance is created according to the obtained manufacturing information and information schema (wipML). The information about manufacturing BOM, which can be captured by EAs, is included in the new instance. A binding model is employed to link the information nodes of manufacturing BOM to related EAs. At the third stage, critical event (CE) structure is used to process the captured manufacturing information from EAs during manufacturing execution process. To meet the requirements of each stage, two components are designed in the PEMA. They are data source service and WIP tracking and tracing, respectively. They are described in detail as follows.

1. Data source service

   Data source service is used to offer a mechanism for the information sharing and integration between EISs and manufacturing execution level. However, the heterogeneity of EISs proposes high challenge to share and integrate the information between these manufacturing elements. To address this problem, Business to Manufacturing Markup Language (B2MML) is employed here to provide standard information schemas for EISs and manufacturing execution level.

   Generally, parameters of data source of the EISs that decision makers want to know or acquire are input in the data source service, and the shared and integrated information which is based on B2MML schemas is the output.

2. WIP tracking and tracing
This component is used to configure distributed EAs based on the relationships between them and to obtain real-time information of WIP in the manufacturing shop floor.
CE structure is adopted in this component. On one hand, it aims to get more useful and actionable information based on the initial information in manufacturing shop floor. On the other hand, it is used as a tool for controlling the event-driven information systems. After getting real-time production information, an instance is created. According to created instance, CE is formed. Then the real-time information of CE can be obtained. Based on the real-time information of CE, users can monitor and control the manufacturing process in the shop floor.

8.8 GA-BASED PRODUCTION SCHEDULING ALGORITHM

This section proposes the GA-based production scheduling algorithm. The GA is a metaheuristic inspired by the process of natural selection that belongs to the larger class of evolutionary algorithms, and it is commonly used to generate high-quality solutions to optimization problems. Three elements are important in the algorithm, namely, gene and chromosome, fitness function, and iteration strategy.

1. Gene and chromosome
At each generation, every chromosome represents a plan for the given production schedule problem. Before the GA can be run, a suitable encoding schema of the problem must be set. The goal is to encode a set of genes (parameters) and to join them together in order to form a chromosome.
In this chapter, the genes and chromosomes are constructed according to the integer-based pattern. An integer i ($1 \le i \le n$) is used to form a gene, where, n is the total number of tasks. The integer will be repeated N_i times for task (i) in the chromosome, and the occurrence order of the integer "i" reflects the processing order. By queuing the different genes and connected by char "-," the chromosome is formed. Thus, the length of the chromosome is determined by the total number of the processes of all the tasks.
2. Fitness function
A fitness function is set to assign a figure of merit to each chromosome. In this chapter, the objective function in Section 8.6 is applied to be the fitness function to estimate the chromosome.
3. Evolution strategy
During each iteration, parents must be selected to generate new offspring. Generally, parents are randomly selected from the population, using a scheme which is created according to the fitness function. Three kinds of evolution mechanisms, that is, selection, crossover, and mutation, are used in GA. Normally, the mechanism "selection" is easy and has well-behaved offspring compared with the other two. Thus, only crossover and mutation mechanisms are presented.

Crossover mechanism is applied to create offspring by recombining the randomly selected parents. Here, we use the multipoint crossover mechanism to change the task order without violating the validity of the produced chromosome. The proposed crossover mechanism contains five main steps.

Step 1: Randomly select two parents, for example, P_1: (3-1-3-2-1-2); P_2: (1-2-3-1-2-3).

Step 2: Choose two positions to generate two crossover sections for the two parents randomly. Such as, if points 2 and 3 are chosen, the selected crossover sections are (1-3) in P_1 and (2-3) in P_2.

Step 3: For each parent, adopt character "0" to substitute the genes in the crossover section of another parent. For example, for P_1, use character "0" to substitute the genes in the crossover section of P_2, that is, the genes "2" and "3." Thus, a new parent P_1': (0-1-3-0-1-2) can be acquired.

Step 4: Move the characters "0," so that they can arrive the cross-section. For example, after this step, P_1' (0-1-3-0-1-2) will be converted to P_1'' (1-0-0-3-1-2).

Step 5: Replace the characters "0" with the crossover section of another parent to create offspring so that the characters "0" in P_1'' (1-0-0-3-1-2) will be substituted by the crossover section of P_2, that is, (2-3). Thus, the new offspring (1-2-3-3-1-2) is created.

Mutation mechanism aims to keep the variety in each population. Only a single chromosome can be altered in the mutation mechanism; the offspring will be generated by changing one or more genes. Contrast to the normal mutation mechanism, an original mutation mechanism is provided to avoid the absence of good results. A performance function is designed to assess each gene of the chromosome, and two points of the inferior genes are randomly selected as the mutation points. The performance function is given in Eq. (8.6).

$$PV_i = \frac{\left(Bt_i^j - Ct_{i-1}^j\right) + \left(Bt_{i+1}^j - Ct_i^j\right)}{Ct_k^j - Bt_0^j} \tag{8.6}$$

where, P_i denotes the performance of gene (i), j represents the equipment ID, k denotes the total number of genes allocated to equipment (j), Bt_i^j means the begin time of gene (i) assigned at equipment (j), Ct_i^j represents the completed time of gene (i) at equipment (j), Bt_{i+1}^j and Ct_{i-1}^j represent the begin time of the later process and the completed time of prior process of gene (i) assigned at equipment (j). Clearly, the lesser the value of P_i, the better the performance of gene (i).

The presented mutation mechanism contains four main steps, as follows.

Step 1: Choose a parent P_1, and use Eq. (8.7) to calculate the performance value for each gene of P_1. And store the values in a matrix

$$M = \{(g_1, pv_1), \ldots, (g_i, pv_i), \ldots, (g_m, pv_m)\} \tag{8.7}$$

where g_i represents the jth gene of P_1 and pv_i means the performance value.

Step 2: Permute M based on the rising value of pv_i and obtain $M^{'}$. In $M^{'}$, the later positions mean the related genes have worse performance. Let $p^{'}$ be a gauge for evaluating the performance of genes. If $pv_i > p^{'}$, the corresponding gene g_i in P_1 is inferior, and the inferior genes can be stored as $IG = \{g_1, \ldots, g_k, \ldots, g_l\}$.

Step 3: Randomly choose two genes from IG, for example, g_i and g_j $(i \neq j)$. These genes are the substituted genes for P_1.

Step 4: Alter the selected genes (g_i and g_j) of P_1 and obtain a new chromosome. That is the mutated offspring.

REFERENCES

[1] S. Weiming, W. Lihui, H. Qi, Agent-based distributed manufacturing process planning and scheduling: a state-of-the-art survey, IEEE Trans. Syst. Man Cybern. C 36 (4) (2006) 563–577.

[2] F.T.S. Chan, J. Zhang, A multi-agent-based agile shop floor control system, Int. J. Adv. Manuf. Technol. 19 (10) (2002) 764–774.

[3] G.Q. Huang, P.K. Wright, S.T. Newman, Wireless manufacturing: a literature review, recent developments, and case studies, Int. J. Comput. Integr. Manuf. 22 (7) (2009) 579–594.

[4] Y. Zhang, G.Q. Huang, T. Qu, O. Ho, S. Sun, Agent-based smart objects management system for real-time ubiquitous manufacturing, Robot. Comput. Integr. Manuf. 27 (3) (2011) 538–549.

[5] G. Wang, T.N. Wong, X. Wang, An ontology based approach to organize multi-agent assisted supply chain negotiations, Comput. Ind. Eng. 65 (1) (2013) 2–15.

[6] R. Macchiaroli, S. Riemma, A negotiation scheme for autonomous agents in job shop scheduling, Int. J. Comput. Integr. Manuf. 15 (3) (2002) 222–232.

[7] N.K.C. Krothapalli, A.V. Deshmukh, Design of negotiation protocols for multi-agent manufacturing systems, Int. J. Prod. Res. 37 (1999) 1601–1624.

[8] H.S. Nwana, D.T. Ndumu, Introduction to agent technology, BT Technol. J. 14 (4) (1996) 55–67.

[9] R. Sikora, M.J. Shaw, A multi-agent framework for the coordination and integration of information systems, Manage. Sci. 44 (11 Part 2) (1998) S65–S78.

[10] H.Z. Jia, S.K. Ong, J.Y.H. Fuh, Y.F. Zhang, A.Y.C. Nee, An adaptive and upgradable agent-based system for coordinated product development and manufacture, Robot. Comput. Integr. Manuf. 20 (2) (2004) 79–90.

[11] J. Jiao, X. You, A. Kumar, An agent-based framework for collaborative negotiation in the global manufacturing supply chain network, Robot. Comput. Integr. Manuf. 22 (3) (2006) 239–255.

[12] Y. Zhang, T. Qu, O.K. Ho, G.Q. Huang, Agent-based smart gateway for RFID-enabled real-time wireless manufacturing, Int. J. Prod. Res. 49 (5) (2011) 1337–1352.

[13] A.M. Farid, L. Ribeiro, An axiomatic design of a multiagent reconfigurable mechatronic system architecture, IEEE Trans. Ind. Inform. 11 (5) (2015) 1142–1155.

[14] A.J.C. Trappey, T.H. Lu, L.D. Fu, Development of an intelligent agent system for collaborative mold production with RFID technology, Robot. Comput. Integr. Manuf. 25 (1) (2009) 42–56.

[15] S. Kumari, A. Singh, N. Mishra, J.A. Garza-Reyes, A multi-agent architecture for outsourcing SMEs manufacturing supply chain, Robot. Comput. Integr. Manuf. 36 (2014) 36–44.
[16] X. Wang, H. Wang, C. Qi, Multi-agent reinforcement learning based maintenance policy for a resource constrained flow line system, J. Intell. Manuf. 27 (2) (2016) 325–333.
[17] B. Vogel-Heuser, C. Diedrich, D. Pantförder, P. Göhner, Coupling heterogeneous production systems by a multi-agent based cyber-physical production system, in: Proceedings—2014 12th IEEE International Conference on Industrial Informatics, INDIN 2014, 2014, pp. 713–719.
[18] V.K. Manupati, G.D. Putnik, M.K. Tiwari, P. Ávila, M. Cunha, Integration of process planning and scheduling using mobile-agent based approach in a networked manufacturing environment, Comput. Ind. Eng. 94 (2016) 63–73.
[19] M.C. Vidoni, A.R. Vecchietti, An intelligent agent for ERP's data structure analysis based on ANSI/ISA-95 standard, Comput. Ind. 73 (2015) 39–50.
[20] M.B. Ayhan, M.E. Aydin, E. Öztemel, A multi-agent based approach for change management in manufacturing enterprises, J. Intell. Manuf. 26 (2013) 1–14.
[21] L.A. Bearzotti, E. Salomone, O.J. Chiotti, An autonomous multi-agent approach to supply chain event management, Int. J. Prod. Econ. 135 (1) (2012) 468–478.
[22] R. Ruiz, C. Maroto, J. Alcaraz, Solving the flowshop scheduling problem with sequence dependent setup times using advanced metaheuristics, Eur. J. Oper. Res. 165 (1) (2005) 34–54.
[23] D.L. Yang, S.J. Yang, Unrelated parallel-machine scheduling problems with multiple rate-modifying activities, Inf. Sci. 235 (2013) 280–286.
[24] W.H. Kuo, C.J. Hsu, D.L. Yang, Some unrelated parallel machine scheduling problems with past-sequence-dependent setup time and learning effects, Comput. Ind. Eng. 61 (1) (2011) 179–183.
[25] Z.X. Guo, W.K. Wong, S.Y.S. Leung, J.T. Fan, S.F. Chan, Genetic optimization of order scheduling with multiple uncertainties, Expert Syst. Appl. 35 (4) (2008) 1788–1801.
[26] E.H. Aghezzaf, Production planning and warehouse management in supply networks with inter-facility mold transfers, Eur. J. Oper. Res. 182 (3) (2007) 1122–1139.
[27] A. Babayan, D. He, Solving the n-job 3-stage flexible flowshop scheduling problem using an agent-based approach, Int. J. Prod. Res. 42 (4) (2004) 777–799.
[28] K. Fang, N. Uhan, F. Zhao, J.W. Sutherland, A new approach to scheduling in manufacturing for power consumption and carbon footprint reduction, J. Manuf. Syst. 30 (4) (2011) 234–240.
[29] G.-C. Luh, C.-H. Chueh, A multi-modal immune algorithm for the job-shop scheduling problem, Inf. Sci. 179 (10) (2009) 1516–1532.
[30] S. Giordani, M. Lujak, F. Martinelli, A distributed multi-agent production planning and scheduling framework for mobile robots, Comput. Ind. Eng. 64 (1) (2013) 19–30.
[31] J. Fan, X. Lu, P. Liu, Integrated scheduling of production and delivery on a single machine with availability constraint, Theor. Comput. Sci. 562 (C) (2015) 581–589.
[32] M.S. Shishvan, J. Sattarvand, Long term production planning of open pit mines by ant colony optimization, Eur. J. Oper. Res. 240 (3) (2014) 825–836.
[33] B.-Y. Cheng, J.Y.-T. Leung, K. Li, Integrated scheduling of production and distribution to minimize total cost using an improved ant colony optimization method, Comput. Ind. Eng. 83 (2015) 217–225.
[34] J.A. Arauzo, J.M. Galán, J. Pajares, A. López-Paredes, Multi-agent technology for scheduling and control projects in multi-project environments. An auction based approach, Intel. Artif. 13 (42) (2009) 12–20.
[35] J.H. Anderson, J.M. Calandrino, Parallel real-time task scheduling on multicore platforms, in: Proceedings—Real-Time Systems Symposium, 2006, pp. 89–100.

[36] P. Tabuada, Event-triggered real-time scheduling of stabilizing control tasks, IEEE Trans. Automat. Contr. 52 (9) (2007) 1680–1685.
[37] A. Subramanian, M.J. Garcia, D.S. Callaway, K. Poolla, P. Varaiya, Real-time scheduling of distributed resources, IEEE Trans. Smart Grid 4 (4) (2013) 2122–2130.
[38] N. Buyurgan, C. Saygin, Application of the analytical hierarchy process for real-time scheduling and part routing in advanced manufacturing systems, J. Manuf. Syst. 27 (3) (2008) 101–110.
[39] Y. Yan, Z. Wang, A two-layer dynamic scheduling method for minimising the earliness and tardiness of a re-entrant production line, Int. J. Prod. Res. 50 (2) (2012) 499–515.
[40] J. Du, J. Park, I. Harjunkoski, M. Baldea, A time scale-bridging approach for integrating production scheduling and process control, Comput. Chem. Eng. 79 (2015) 59–69.
[41] S. Lee, V.V. Prabhu, A dynamic algorithm for distributed feedback control for manufacturing production, capacity, and maintenance, IEEE Trans. Autom. Sci. Eng. 12 (2) (2015) 628–641.
[42] H. Luo, J. Fang, G.Q. Huang, Real-time scheduling for hybrid flowshop in ubiquitous manufacturing environment, Comput. Ind. Eng. 84 (2015) 12–23.
[43] H.S. Choi, J.S. Kim, D.H. Lee, Real-time scheduling for reentrant hybrid flow shops: a decision tree based mechanism and its application to a TFT-LCD line, Expert Syst. Appl. 38 (4) (2011) 3514–3521.
[44] Y. Zhang, G. Zhang, J. Wang, S. Sun, S. Si, T. Yang, Real-time information capturing and integration framework of the Internet of manufacturing things, Int. J. Comput. Integr. Manuf. 3052 (December) (2014) 1–12.
[45] L. Da Xu, W. He, S. Li, Internet of things in industries: a survey, IEEE Trans. Ind. Inform. 10 (4) (2014) 2233–2243.
[46] G.Q. Huang, Y.F. Zhang, P.Y. Jiang, RFID-based wireless manufacturing for walking-worker assembly islands with fixed-position layouts, Robot. Comput. Integr. Manuf. 23 (4) (2007) 469–477.
[47] H. Wang, S. Chen, Y. Xie, An RFID-based digital warehouse management system in the tobacco industry: a case study, Int. J. Prod. Res. 48 (9) (2010) 2513–2548.
[48] M. Grüninger, S. Shapiro, M.S. Fox, H. Weppner, Combining RFID with ontologies to create smart objects, Int. J. Prod. Res. 48 (9) (2010) 2633–2654.
[49] D.F. Xu, Q. Li, H.B. Jun, J. Browne, Y.L. Chen, D. Kiritsis, Modelling for product information tracking and feedback via wireless technology in closed-loop supply chains, Int. J. Comput. Integr. Manuf. 22 (7) (2009) 648–670.
[50] Y.C. Lin, W.F. Cheung, F.C. Siao, Developing mobile 2D barcode/RFID-based maintenance management system, Autom. Constr. 37 (2014) 110–121.
[51] S.K. Al Subramaniam, S.H. Binti Husin, Y. Binti Yusop, A.H. Bin Hamidon, Real time shop floor monitoring system for a better production line management, in: 2007 Asia-Pacific Conference on Applied Electromagnetics Proceedings, APACE2007, 2007.

Chapter 9

IoT-MS Prototype System

9.1 CONFIGURATION OF A SMART SHOP FLOOR

An Internet of things (IoT)-enabled environment is essential to the implementation of Internet of things based manufacturing system (IoT-MS). In this chapter, to test the effectiveness of the proposed system, some production tasks are formed first. Then, a layout of shop floor is provided, followed by the deployment of sensors and the configuration of machines. All these operations build up the hardware foundations of IoT-MS.

9.1.1 Formation of the Production Task

For simplicity of understanding but without losing generality of principle, we collected some data from a mechanical manufacturer and modeled their processing route with Tecnomatix Plant Simulation.

Tecnomatix Plant Simulation is an industrial solution provided by Siemens Product Lifecycle Management Software Inc. which can be used to model the sophisticated manufacturing processes, material flows, shop-floor layouts, etc. Its features of object-oriented architecture, discrete event simulation, and three-dimensional display of models allow users to build realistic manufacturing models precisely and based on real-life data [1].

Their main business is to produce different types of speed transmission. The major components of one particular product or its bill of materials (BOM) is shown in Fig. 9.1.

For further study, only part of their manufacturing resource model was established in Fig. 9.2 by using the previously mentioned software. The actual manufacturing route is more complicated because it involves multiple products. To increase the usage of machines, some stations were designed flexible and were occupied by several production lines. We used this software to identify the critical production routes and to simplify this particular manufacturing system, so as to demonstrate the prototype system more clearly. To highlight the operating processes of our prototype system, only key processes will be involved in the following discussion.

The detailed information of the task is described as follows. The structure of the task can be seen in Fig. 9.3. Four separate products are included in this particular order. Each product needs to be assembled from different parts.

Optimization of Manufacturing Systems Using the Internet of Things
http://dx.doi.org/10.1016/B978-0-12-809910-0.00009-3

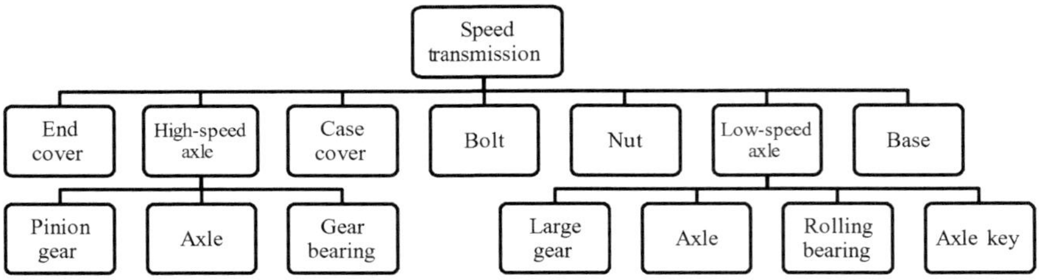

FIGURE 9.1 Bill of materials (BOM) of the product.

To specify, "Product 1" requires operation "Assembly 1," which is to assemble "Part 11" and "Part 12" together. Each part needs to be manufactured through multiple processes, which are represented by gray blocks in Fig. 9.3. The scheduled processing time of each process is listed in Table 9.1.

According to the theories discussed in Section 5.7.1, qualified manufacturing cell (MC) can undertake the decomposed subtle tasks which are above the dotted line. For example, the operation of "Assembly 2," "Part 31," etc. can be done by a combination of machines or at a combination of several stations. The decomposed subtle tasks, which are below the dotted line, can be assigned to certain machines in MCs. These decomposed subtle tasks are usually manufacturing processes.

9.1.2 Layout of the Shop Floor

Again, for simplicity of understanding, we made some modifications to the actual layouts of the manufacturer and formed the discrete manufacturing environment. The layout of the case scenario is shown in Fig. 9.4.

In this typical manufacturing system, the fundamental manufacturing resources are designed as follows.

There are 12 manufacturing machines and 2 assembly stations, hereinafter referred to as manufacturing stations. These manufacturing machines include five lathes, four milling machines, two gear-hobbing machines, and one numerical controlled processing center. For each manufacturing station, there are two buffers, acting as a material entrance and an exit, respectively. Materials will be sent to the entrance, waiting to be processed when the machine is available; and will be put to the exit, waiting to be carried by some vehicles. Manufacturing tools, fixtures, measuring tools, and backup parts are stored in the warehouses. Some raw materials, work-in-progress parts, and finished products are also temporarily stored there. Multiple forklifts and handcarts move along the paths to fulfill the material handling tasks. The detailed information of each objects in the shop floor are listed in Table 9.2.

9.1.3 Deployment of Hardware Devices

Large amount of data need to be collected to achieve the proactive sensing, real-time monitoring, and intelligent decision making during production. To capture

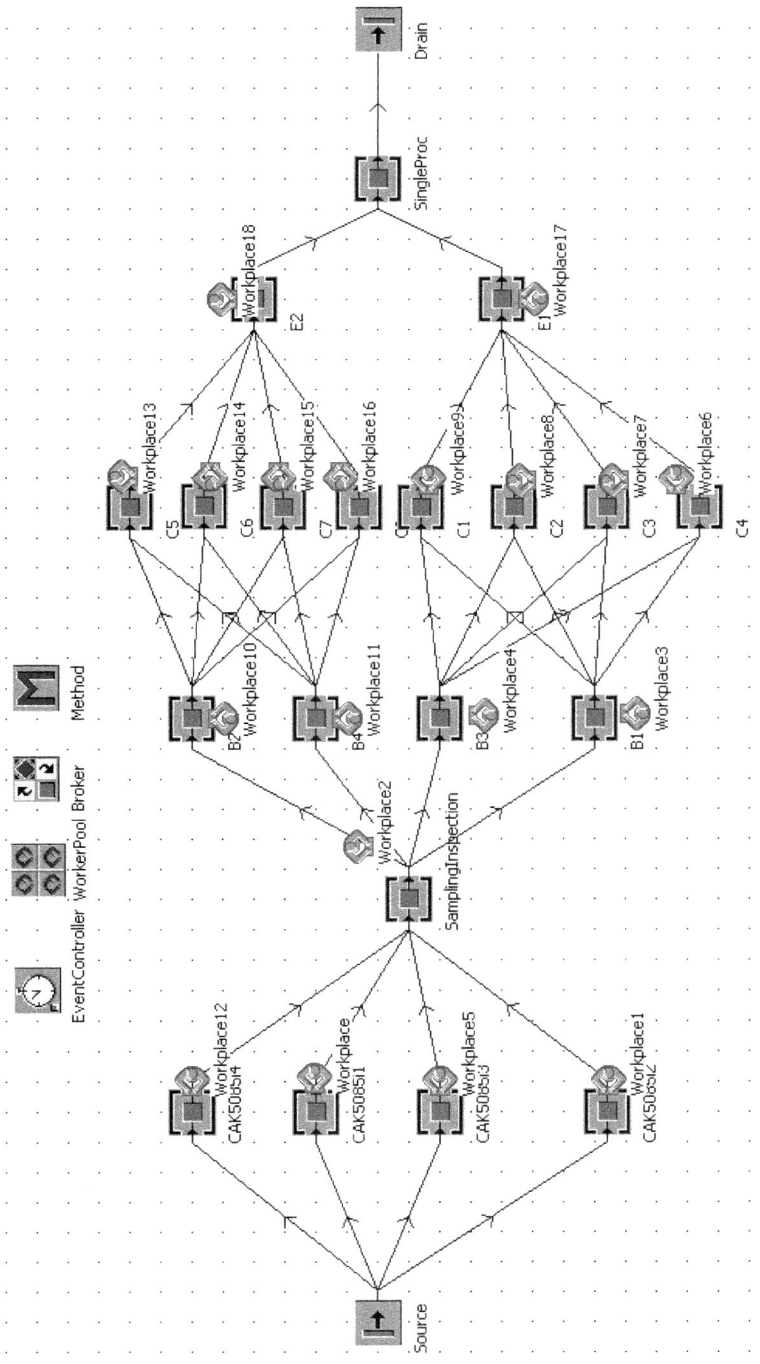

FIGURE 9.2 Modeling of the manufacturing process.

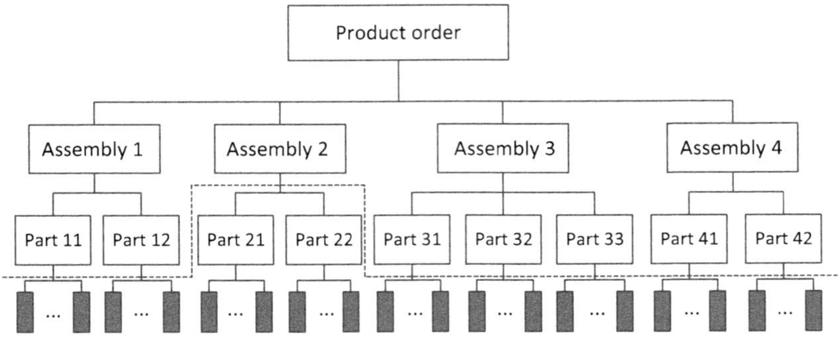

FIGURE 9.3 The structure of the task.

TABLE 9.1 The Production Schedule of Processing Tasks

Different levels of processing tasks			Deadline
Product order			550
	Assembly 1		450
		Part 11	185
		Part 12	170
	Assembly 2		380
		Part 21	—
		Part 22	—
	Assembly 3		430
		Part 31	180
		Part 32	180
		Part 33	175
	Assembly 4		400
		Part 41	150
		Part 42	200

the data related to production progress, RFID readers are set up in the entrances of material, exits of material, areas of work-in-process (WIP) parts, and processing areas, respectively. These RFID readers can be used to track the material flow and to get the location information of operators in real time. Such data are important inputs of dynamic allocation of resources, production rescheduling, etc. As shown in Fig. 9.5, RFID readers are mounted in different locations. RFID tags are attached to products or held by operators. When a tag is detected by a reader, the rough location of the product or the operator can be determined.

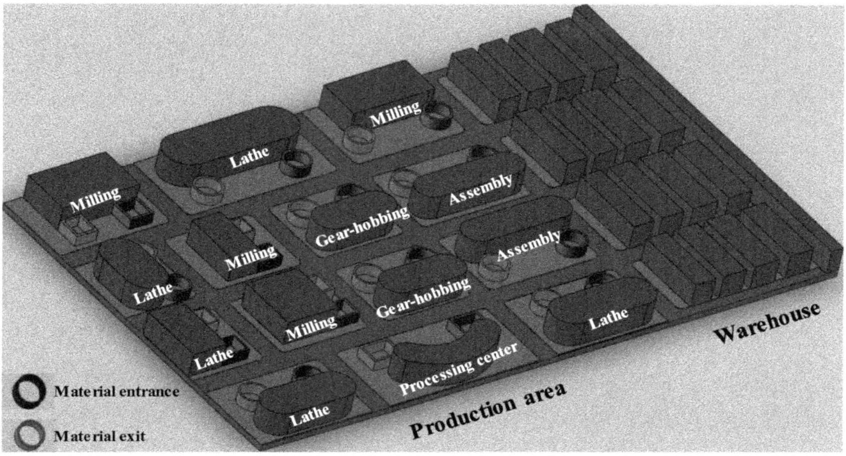

FIGURE 9.4 **Layout of the manufacturing shop floor.**

TABLE 9.2 Detailed Information of the Shop Floor

Name	Type	Quantity
Lathe	Machine/station	5
Milling machine	Machine/station	4
Gear-hobbing machine	Machine/station	2
Numerical controlled processing center	Machine/station	1
Assembly station	Workplace/station	2
Storage rack	Infrastructure	35
Material buffer	Infrastructure	28
Forklift	Vehicle	4
Handcart	Vehicle	6

Similarly, RFID readers and tags are also used in the material handling system. Multiple RFID readers are installed on forklifts or handcarts. Some readers are installed at the bottom of the vehicle, as shown in Fig. 9.6A–B. These readers (or antennas) can identify the tags that are located in some key locations of the shop floor as shown in Fig. 9.6C. Thus, the locations of forklifts or handcarts are available to the material handling system. For such purposes, the narrow-band antennas are applied to limit the sensing ranges and to achieve relatively precise results. Additional RFID readers can be installed on the vehicles to achieve different goals. For example, some readers are installed where the products will be placed. These readers can capture the data of the carried products and locate

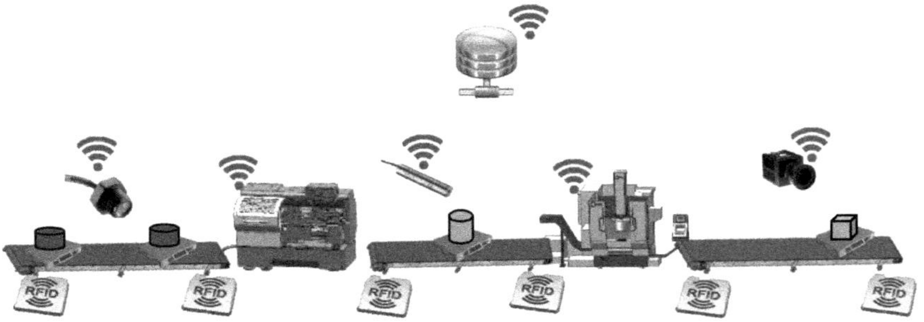

FIGURE 9.5 **Deployment of the material tracking system.**

the exact position of a bay inside a shelf as shown in Fig. 9.6A–B. Wide-band antennas can be applied here because they can sense for wider ranges. Products then do not necessarily need to be placed very close to the antennas in order to be perceived.

The "multitype data capturing device" in Fig. 9.6 refers to the real-time and multiple-source manufacturing information sensing system (MISS), which has been discussed in detail in Chapter 3. As shown in Fig. 9.7, by applying the technologies described in former chapters, the traditional manufacturing machines are now equipped with a processing unit, a communication module, some embedded sensors, etc.; all provided by MISS.

9.2 THE FRAMEWORK OF THE PROTOTYPE SYSTEM

9.2.1 System Architecture

The overall framework of the prototype system is shown in Fig. 9.8. The whole system is driven by the real-time data captured from multiply sensors. The cloud computing based manufacturing resources configuration method is responsible to pair tasks and machines with suitable capabilities. The IoT-enabled smart trolley handles the moving tasks within the shop floor. There are also a number of functional modules, including real-time and multisource MISS, IoT-enabled smart stations, real-time key production performance analysis method, and real-time information driven production scheduling system. These modules will analyze the captured data and provide feedback to machines or guidance for operators.

9.2.2 Information Model

The information model of the prototype system is shown in Fig. 9.9. This model will guide the design of information databases used in this system. Basically, there are 14 data sheets, including the real-time information of equipment, the real-time information of the shop floor, the real-time information of WIPs, real-time information of vehicles, history of equipment, history of the shop floor,

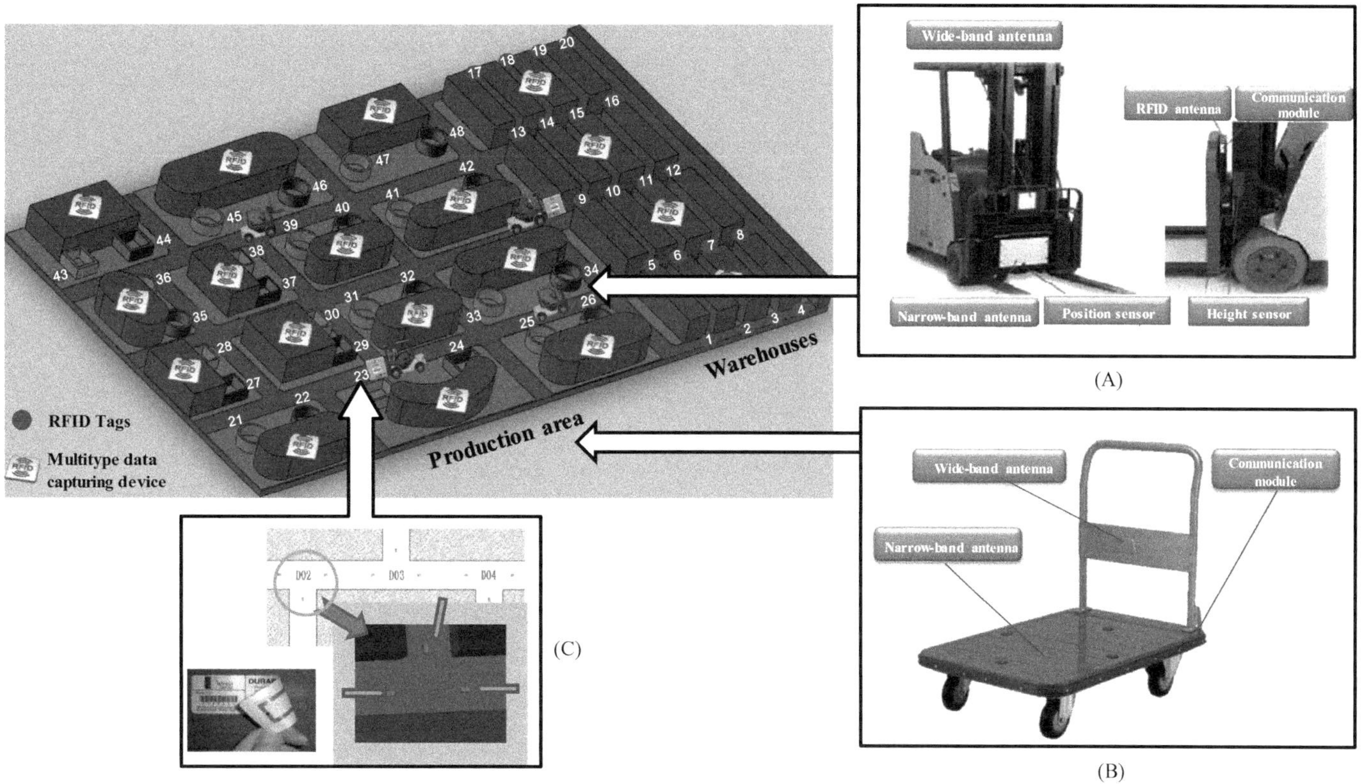

FIGURE 9.6 The configuration of a smart shop floor. (A) The deployment of the forklift. (B) The deployment of the handcart. (C) The location of RFID tags.

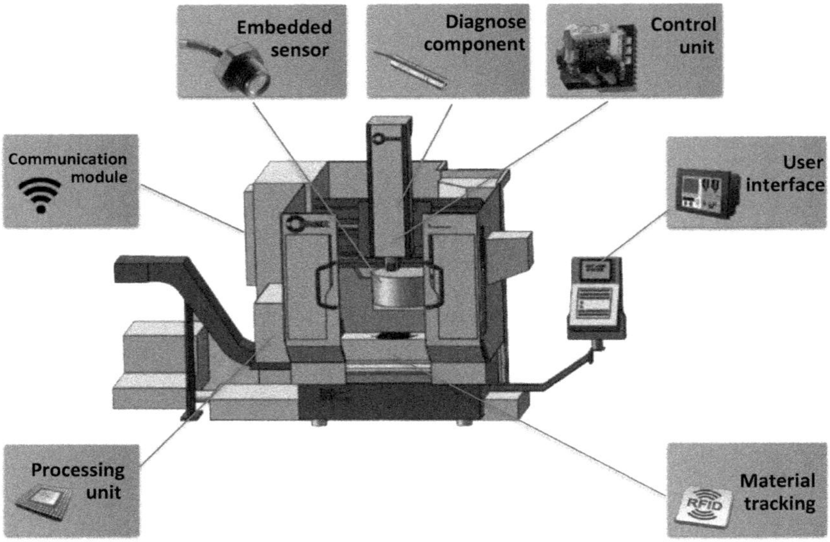

FIGURE 9.7 **Component of a smart machine.**

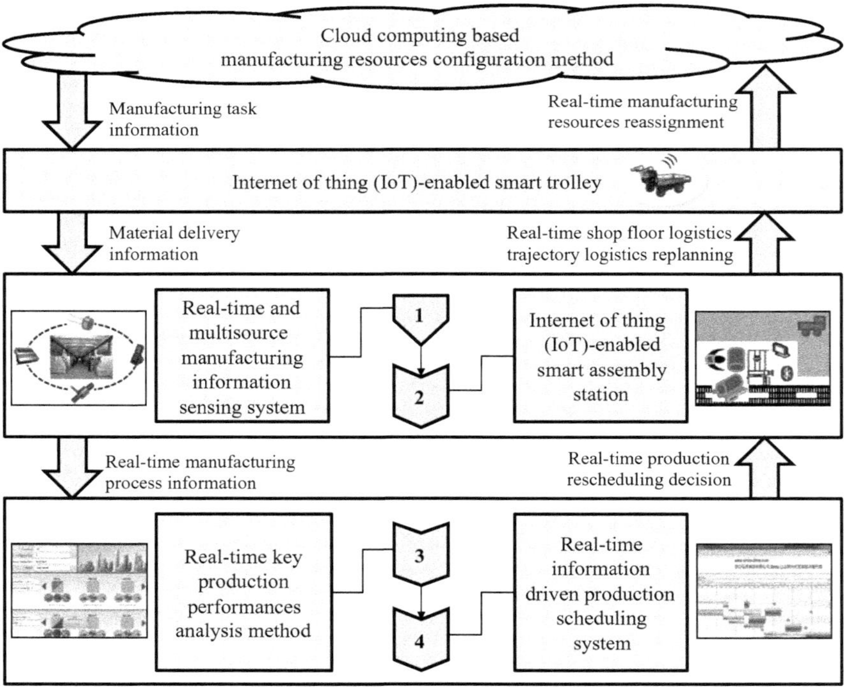

FIGURE 9.8 **Overall framework of the prototype system.**

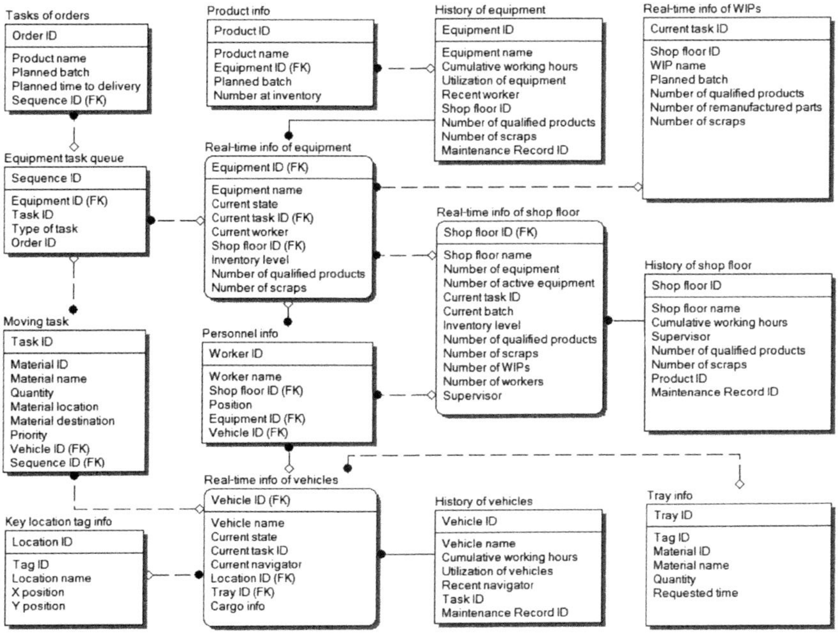

FIGURE 9.9 The information model of the prototype system.

history of vehicles, tray information, personnel information, product information, the tag information of key positions, the moving tasks of vehicles, the task queue of equipment, and the tasks from the orders. Possible fields are also shown in Fig. 9.9. Those fields listed on the top of each sheet are the primary keys, and those marked with "(FK)" indicates that the field is a foreign key of another data sheet.

9.3 THE LOGICAL FLOW OF THE PROTOTYPE SYSTEM

The logical flow of the prototype system can be seen in Fig. 9.10. First, the IoT devices will be set up according to Section 9.1. The multitype and real-time manufacturing information can be accessible from the constructed environment. After the set-up, the system will work under the following steps.

Step 1: Upon receiving the tasks from customers, the cloud computing based manufacturing resources configuration method will try to decompose the tasks to manufacturing process level. According to the current status of machines and the requirements of tasks, the tasks will be coupled to the machines with proper capabilities.

Step 2: The paring results will be further processed, since different processing tasks may be assigned to the same machine at the same time. The

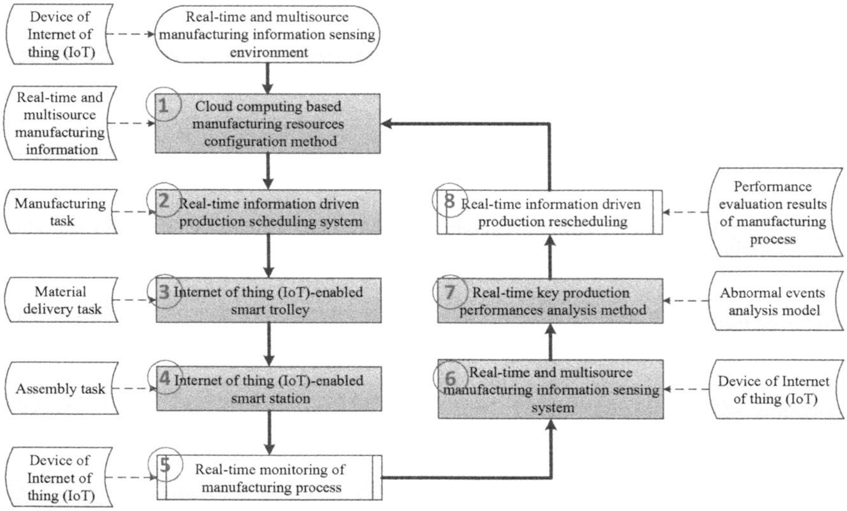

FIGURE 9.10 Workflow of the prototype system.

real-time information driven production scheduling system will find out the optimal production plan, which will be executed soon.

Step 3: Since the production plans have been decided, the raw materials need to be brought to the corresponding stations to initialize production. The IoT-enabled smart trolley is responsible to plan the routes for vehicles and guide them to the correct sites with highest efficiency.

Step 4: After obtaining the correct materials, manufacturing stations will start production. The IoT-enabled smart station will provide guidance to the operators if necessary, and will update the task queue according to real-time information. The station will also communicate with upstream and downstream stations on their production progress, so that any potential delays or other manufacturing exceptions can be identified rapidly for early preparation and early solutions.

Step 5: During the manufacturing execution stage, the real-time monitoring module will keep monitoring the manufacturing processes in case of exceptions.

Step 6: The real-time and multisource MISS will collect the necessary data for further analysis when monitoring.

Step 7: The key production performance analysis method will try to determine the critical events during production and find out the potential causes of manufacturing exceptions.

Step 8: The rescheduling model will rearrange the production plans according to the information from performance analysis. This process continues until the end of production and the arrival of new tasks.

9.4 TASK DRIVEN MANUFACTURING RESOURCE CONFIGURATION MODULE

A typical production task will specify what is to be produced, what are the quality requirements, what is the production batch, how much is the cost, and what is the delivery time of the final products. Since the deadline of the task is available to IoT-MS, the expected finishing time for all operations can be obtained basically according to the manufacturing routes. Then, the deadline for each subtle task is listed in Table 9.1. The resource configuration module includes two phases, namely the optimal configuration of the MC and the cloud manufacturing service (CMS) [2].

9.4.1 Phase 1: MC Optimal Configuration

Let $T = \{ST_i; i = 1, 2, 3, \ldots, 8\}$. Suppose T represents the overall task requested from the order, which can be decomposed into eight subtle tasks undertaken by MCs as shown in Fig. 9.3. Eight candidate sets of MCs (MCCS) are formed accordingly. For example, in this case, there are seven potential service providers for Assembly 2, which can be represented as $MCCS_3 = \{MC_3^1, MC_3^2, MC_3^3, MC_3^4, MC_3^5, MC_3^6, MC_3^7\}$. The evaluation indicators for each candidate service are shown in Table 9.3.

These candidate services are assessed by adopting the evaluation method based on GRA illustrated in Section 5.7.2. The detailed calculations are included as follows.

Step 1: The optimal indicator sequence is obtained according to (5.8).

$$S^* = (550, 380, 0, 85, 9.2, 39)$$

TABLE 9.3 The Evaluation Indicators for Candidate Services of Assembly 2

Candidate services	Evaluation criteria					
	C	DT	dt	R	C_r	E
MC_3^1	590	405	25	75	8.3	42
MC_3^2	630	370	0	83	8.8	43
MC_3^3	580	400	20	75	8.4	39
MC_3^4	550	380	0	79	8.1	43
MC_3^5	565	370	0	85	9.2	42
MC_3^6	610	375	0	80	7.9	40
MC_3^7	615	360	0	83	8.5	47

Step 2: The normalized evaluation matrix can be derived according to (5.9–5.11).

$$S_N = \begin{bmatrix} 0.5 & 0 & 0 & 0 & 0.307692 & 0.625 \\ 0 & 0.777778 & 1 & 0.8 & 0.692308 & 0.5 \\ 0.625 & 0.111111 & 0.2 & 0 & 0.384615 & 1 \\ 1 & 0.555556 & 1 & 0.4 & 0.153846 & 0.5 \\ 0.8125 & 0.777778 & 1 & 1 & 1 & 0.625 \\ 0.25 & 0.666667 & 1 & 0.5 & 0 & 0.875 \\ 0.1875 & 1 & 1 & 0.8 & 0.461538 & 0 \end{bmatrix}$$

Step 3: The relational coefficient matrix is calculated by adopting (5.12).

$$E = \begin{bmatrix} 0.5 & 0.333333 & 0.333333 & 0.333333 & 0.419355 & 0.571429 \\ 0.333333 & 0.555556 & 1 & 0.714286 & 0.619048 & 0.5 \\ 0.571429 & 0.384615 & 0.384615 & 0.333333 & 0.448276 & 1 \\ 1 & 1 & 1 & 0.454545 & 0.371429 & 0.5 \\ 0.727273 & 0.555556 & 1 & 1 & 1 & 0.571429 \\ 0.4 & 0.714286 & 1 & 0.5 & 0.333333 & 0.8 \\ 0.380952 & 0.384615 & 1 & 0.714286 & 0.481481 & 0.333333 \end{bmatrix}$$

Step 4: According to Eq. (5.13), the grey relational degree of each candidate service is derived.

The weights for each evaluation indicator, which are determined by the analytic hierarchy process (AHP) in this case, are denoted as w = (0.186, 0.195, 0.205, 0.147, 0.138, 0.129)T. Then, the comprehensive evaluation matrix is obtained as:

$$R = EW = [0.407, 0.630, 0.499, 0.769, 0.807, 0.641, 0.565].$$

Similarly, by evaluating other candidate services based on the previously mentioned method, their corresponding comprehensive evaluation matrices can be achieved. Here, the top three services are selected in each descending queue in terms of respective relational degrees. The evaluation indicators for all candidate services are shown in Table 9.4.

As shown in Table 9.4, a total of 3^8 service compositions can be generated. By calculating the relational degrees of all possible compositions, the highest result is 0.7959, and the corresponding optimal composition for T is $\{MC_1^3, MC_2^4, MC_3^5, MC_4^5, MC_5^6, MC_6^4, MC_7^7, MC_8^9\}$.

9.4.2 Phase 2: CMS Optimal Configuration

Based on the optimal service composition calculated earlier, all eight subtle tasks are assigned to corresponding MCs. For example, Part 41 is undertaken by MC_7^7. Then the task will be decomposed into six process-level operations

TABLE 9.4 The Evaluation Indicators for the Candidate Services of *T*

Candidate services	Evaluation criteria					
	C	DT	dt	R	C_r	E
MC_1^3	185	185	0	0.95	9.3	19.6
MC_1^6	180	170	0	0.89	7.9	14
MC_1^2	190	170	0	0.85	8.8	16
MC_2^1	295	170	0	0.91	9.1	23.2
MC_2^4	250	160	0	0.88	7.8	21.6
MC_2^3	275	165	0	0.89	8.2	26
MC_3^5	565	370	0	0.85	9.2	42
MC_3^4	550	380	0	0.79	8.1	43
MC_3^6	610	375	0	0.80	7.9	40
MC_4^5	185	160	0	0.92	8.7	32
MC_4^7	245	180	0	0.88	8.8	34
MC_4^8	189	165	0	0.89	9	36
MC_5^6	395	175	0	0.94	8.9	27.6
MC_5^2	410	180	0	0.88	7.3	26
MC_5^3	400	160	0	0.82	9	28
MC_6^6	395	175	0	0.95	8.9	39.5
MC_6^9	375	175	0	0.84	8.8	35
MC_6^4	360	155	0	0.88	8.5	36.5
MC_7^7	275	150	0	0.90	7.9	48.5
MC_7^2	320	150	0	0.83	8.8	44
MC_7^8	280	145	0	0.85	8.0	49
MC_8^9	330	190	0	0.85	7.9	26
MC_8^4	360	200	0	0.85	8.1	30
MC_8^7	365	170	0	0.87	8.4	32

in a certain sequence based on manufacturing routes as shown in Fig. 9.11. The deadline of each process is defined in Table 9.5. Related manufacturing machines are pooled into corresponding candidate sets for each subtask. Top K_g services ($K_g = 3$) in each queue are selected to constitute compositions as shown in Table 9.6.

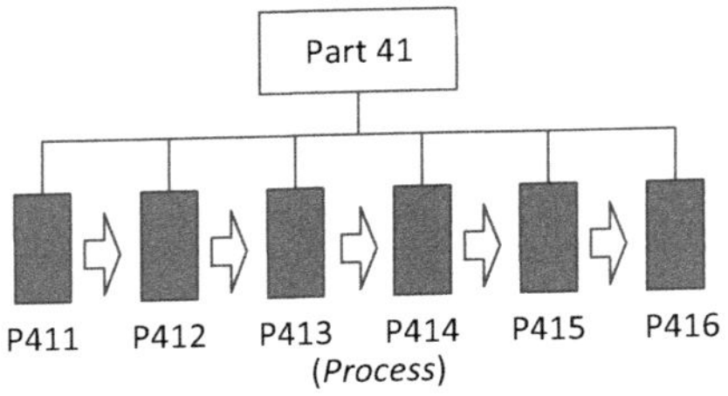

FIGURE 9.11 **The process flow of Part 41.**

TABLE 9.5 The Production Schedule of Part 41

Process					
P411	P412	P413	P414	P415	P416
24	30	20	22	26	28

By calculating the relational degrees of all service compositions, the highest one is achieved as 0.7046. Accordingly, the optimal service composition for Part 41 is $\{ MM_1^6, MM_2^5, MM_3^5, MM_4^7, MM_5^5, MM_6^2 \}$.

9.5 PRODUCTION SCHEDULING/RESCHEDULING MODULE

Based on the discussions in Chapter 8, this section illustrates the real-time production scheduling and rescheduling method by analyzing the formed case.

9.5.1 Quantifying the Tasks

Ten tasks (including assembly, part producing, etc.) and 10 machines (3 lathes, 2 milling machines, 2 gear-hobbing machines, 1 numerical controlled processing center, and 2 assembly stations) are involved in this scheduling problem. Suppose that each manufacturing task has four processes. Table 9.7 shows the requirements of the 10 tasks. Each column represents a different task and the four rows within each column represents four manufacturing processes. The pair of number (x, y) of the jth row and ith column indicates that the process "j" of task "i" needs to be processed on any machines within machine cell "x" and the processing time is "y." For example, the pair (4, 21) of the fifth column and the second row means that the second process of Task 5 need to be processed on any machines of machine cell 4, with 21 units of processing time. In this case, Machine 1, 2, and 3 belongs to MC 1; Machine 4 and 5 belongs to MC 2; Machine 6, 7, and 8 are members of MC 3; The rest of machines are grouped

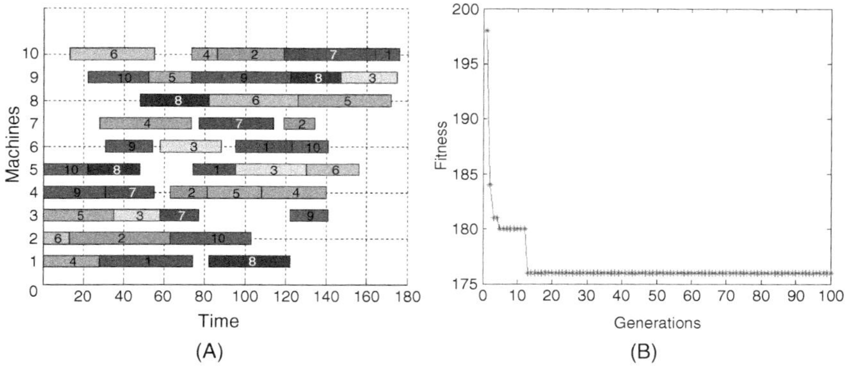

FIGURE 9.12 Scheduling result. (A) The Gantt Chart of the task assignment. (B) The generations and the fitness curve of the designed algorithm.

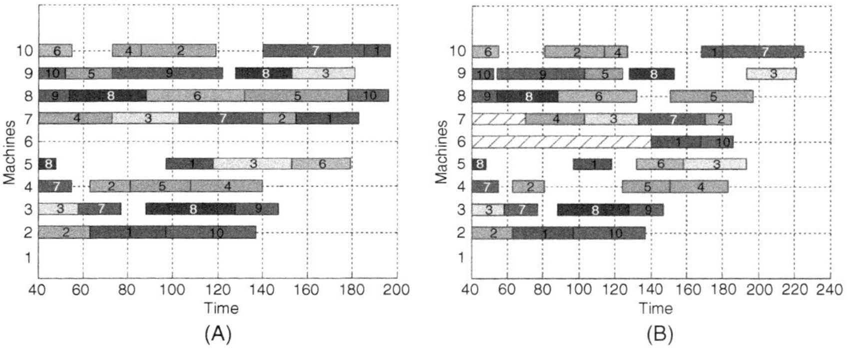

FIGURE 9.13 The rescheduling results under two types of exceptions. (A) Exceptions with unknown recovering time. (B) Exceptions with an exact time window for recovering.

recovery time. Recovering time of the first kind is unknown, while that of the second type is an exact time window. As seen in Fig. 9.13A, the first exception occurred at Machine 1 and Machine 6 at time t (t = 40). In Fig. 9.13B, two kinds of exceptions from Machine 1, Machine 6, and Machine 7 also occur at time t (t = 40). All these exceptions are captured by the PEMA and reported to RSA. For the first kind of exceptions, Machine 1 and Machine 6 in Fig. 9.13A will lose the capability to take any tasks. For the second kind of exceptions, the recovery time window is considered as a load to Machine 6 and Machine 7 in Fig. 9.13B. By running through the first and the second stages, the new scheduling plan is established.

9.6 IoT-ENABLED SMART MATERIAL HANDLING MODULE

9.6.1 Task Description

The processing tasks have been assigned to different stations, so the forklifts, handcarts, or other vehicles need to prepare materials for the stations. In our case, four forklifts are on duty and 15 move tasks are waiting to be finished. The detailed information of vehicles and move tasks are listed in Tables 9.9 and 9.10. PID represents the ID for different locations. The PID of each location is unique. These PIDs are marked on the RFID tags in Fig. 9.6. Then, the distance of each handling task can be easily calculated according to Chapter 6. In the column named Priority, the number "1" represents common tasks, "2" represents important tasks, "3" represents critical tasks, and "4" represents emergencies.

TABLE 9.9 Information of Trolleys

ID	PID (current location)	Maximum usable space	Occupied space
VID_1	10	15	12
VID_2	38	15	8
VID_3	33	15	0
VID_4	26	15	8

TABLE 9.10 Information of the Moving Tasks

ID	PID (from-location)	PID (to-location)	Due time	Priority	Index no.	Product volume
TID_1	5	32	150	1	IID_1	11
TID_2	45	27	140	1	IID_2	3
TID_3	23	35	130	1	IID_3	5
TID_4	11	44	120	1	IID_4	12
TID_5	21	37	110	1	IID5	8
TID_6	16	34	100	1	IID_6	4
TID_7	43	29	90	2	IID_7	10
TID_8	13	24	80	2	IID_8	6
TID_9	25	9	70	2	IID9	9
TID_{10}	39	22	60	2	IID_{10}	7
TID_{11}	28	40	50	3	IID_{11}	5
TID_{12}	31	46	40	3	IID_{12}	7
TID_{13}	7	42	30	3	IID_{13}	10
TID_{14}	30	48	20	4	IID_{14}	4
TID_{15}	41	19	10	4	IID_{15}	6

9.6.2 Calculations for the Moving Tasks

Step 1: Construct real-time information model (V) [4] of trolleys according to the information listed in Table 9.9:

$$V = \begin{bmatrix} VID_1 & 10 & 15 & 12 \\ VID_2 & 38 & 15 & 8 \\ VID_3 & 33 & 15 & 0 \\ VID_4 & 26 & 15 & 8 \end{bmatrix}$$

Step 2: Construct real-time information model (N) of distribution tasks according to the information listed in Table 9.10. Select five tasks to form the candidate task set and construct information model (q) of the candidate task set. Table 9.11 lists the detailed information of the candidate task set.

$$N = \begin{bmatrix} TID_1 & 5 & 32 & 150 & 1 & IID_1 \\ TID_2 & 45 & 27 & 140 & 1 & IID_2 \\ TID_3 & 23 & 35 & 130 & 1 & IID_3 \\ TID_4 & 11 & 44 & 120 & 1 & IID_4 \\ TID_5 & 21 & 37 & 110 & 1 & IID_5 \\ TID_6 & 16 & 34 & 100 & 1 & IID_6 \\ TID_7 & 43 & 29 & 90 & 2 & IID_7 \\ TID_8 & 13 & 24 & 80 & 2 & IID_8 \\ TID_9 & 25 & 9 & 70 & 2 & IID_9 \\ TID_{10} & 39 & 22 & 60 & 2 & IID_{10} \\ TID_{11} & 28 & 40 & 50 & 3 & IID_{11} \\ TID_{12} & 31 & 46 & 40 & 3 & IID_{12} \\ TID_{13} & 7 & 42 & 30 & 3 & IID_{13} \\ TID_{14} & 30 & 48 & 20 & 4 & IID_{14} \\ TID_{15} & 41 & 19 & 10 & 4 & IID_{15} \end{bmatrix}$$

$$q = \begin{bmatrix} TID_{11} & 28 & 40 & 50 & 3 & IID_{11} \\ TID_{12} & 31 & 46 & 40 & 3 & IID_{12} \\ TID_{13} & 7 & 42 & 30 & 3 & IID_{13} \\ TID_{14} & 30 & 48 & 20 & 4 & IID_{14} \\ TID_{15} & 41 & 19 & 10 & 4 & IID_{15} \end{bmatrix}$$

Step 3: It can be inferred from the information model V that Vehicle 3 is idle. Then form the combinations of the moving tasks related with Vehicle 3 according to the rules and methods mentioned in Section 6.6.

Step 4: Construct objective function and select the best combination of moving tasks. The parameters of the model are listed in Table 9.12 and all the parameters of the combinations of moving tasks are listed in Table 9.13.

TABLE 9.11 Information of Candidate Task Set

Code	PID (from-location)	PID (to-location)	Due time	Priority	Index no.	Volume
TID_{11}	28	40	50	3	IID_{11}	5
TID_{12}	31	46	40	3	IID_{12}	7
TID_{13}	7	42	30	3	IID_{13}	10
TID_{14}	30	48	20	4	IID_{14}	4
TID_{15}	41	19	10	4	IID_{15}	6

TABLE 9.12 Parameters of the Model

Parameter	P_0	L_0	U_0	w_p	w_L	w_u
Value	4.833	214	9.75	0.333	0.333	0.333

TABLE 9.13 Parameters of the Combinations of Moving Tasks

Vehicle	Task combination	Priority (*P*)	Distance (*L*)	Volume (*U*)	Value of *f*(*P*, *L*, *U*)
3	TID_{11}	3	141	5	0.801
3	TID_{12}	3	129	7	0.910
3	TID_{13}	3	163	10	0.902
3	TID_{14}	4	153	4	0.787
3	TID_{15}	4	143	6	0.883
3	TID_{11},TID_{12}	6	225	12	1.024
3	TID_{11},TID_{13}	6	255	15	1.088
3	TID_{11},TID_{14}	7	245	9	0.958
3	TID_{11},TID_{15}	7	207	11	1.073
3	TID_{12},TID_{14}	7	173	11	1.136
3	TID_{12},TID_{15}	7	235	13	1.101
3	TID_{13},TID_{14}	7	183	14	1.213
3	TID_{14},TID_{15}	8	229	10	1.066
3	TID_{11},TID_{14},TID_{15}	11	293	15	*1.335*

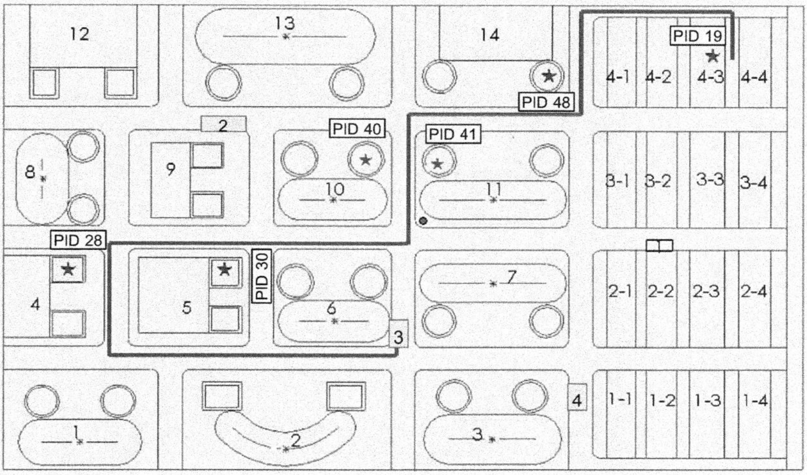

FIGURE 9.14 Routing of the moving task TID_{11}, TID_{14}, and TID_{15}.

We can conclude from Table 9.13 that the combination of moving task TID_{11}, TID_{14}, and TID_{15} has the biggest $f(P, L, U)$ value. Consequently, these three moving tasks will be handled together by Vehicle 3. Fig. 9.14 shows the routing of TID_{11}, TID_{14}, and TID_{15} for Vehicle 3.

9.6.3 User Interfaces of the Prototype System

Figs. 9.15–9.19 demonstrates the flow of operating the software system for intelligent trolley based on real-time data. The flow goes from obtaining a new move task for an idle intelligent trolley. As seen in Fig. 9.15, when an operator approaches the vehicle, the staff information will be sensed and displayed on the upper right corner of the screen. The current status of this vehicle, that is, idle, will be posted to the server.

Based on the previously mentioned discussions, the server will assign the optimal task to this vehicle. In this case, the assigned task is to pick up some parts at site M4, and have it transported to site M14. The current task list and the real-time information are shown on the left side of the screen as in Fig. 9.16. The planned route is also shown on the screen in green arrows. Operators should follow the instructions to reach the site for picking up products.

When the vehicle arrives at the site for picking up, materials that are to be loaded will be listed in the lower left corner of the screen. Operators may click on each items of the list to see a picture of the targeted part, and a diagram of the shelf, clearly indicating where exactly is the part within the racks on this shelf. All parts that have not been loaded will be identified as shown in Fig. 9.17. During the loading of materials, the movement of the materials can be

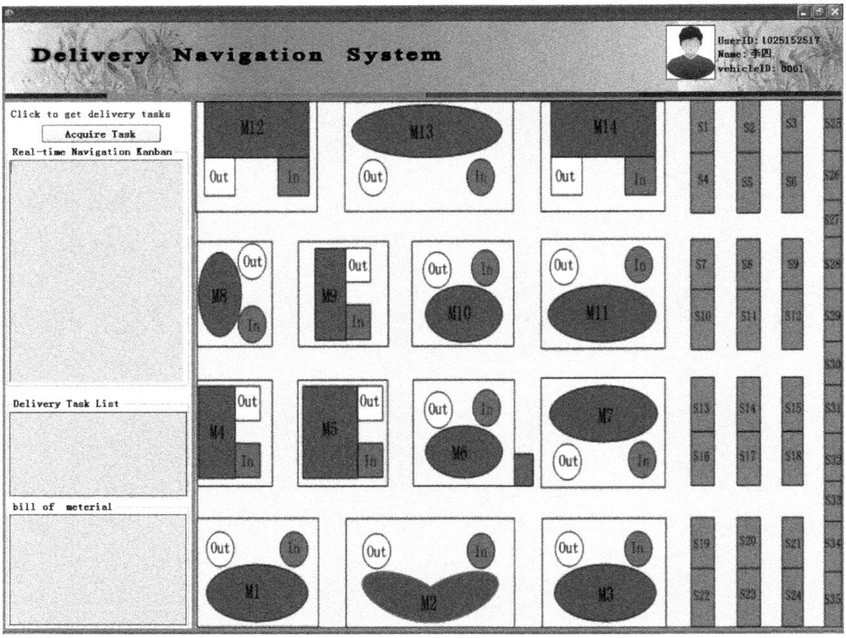

FIGURE 9.15 **Obtaining a new moving task.**

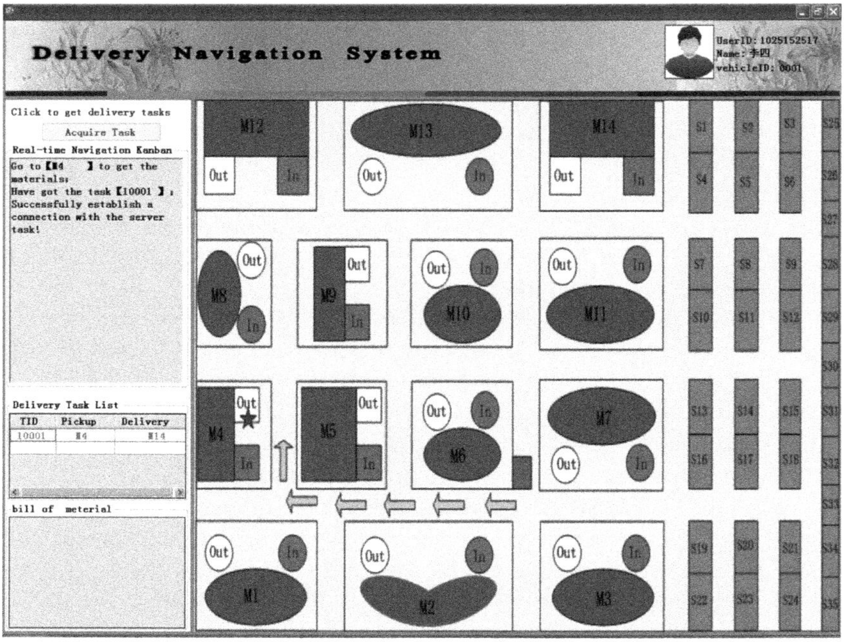

FIGURE 9.16 **Navigating to the pick-up point.**

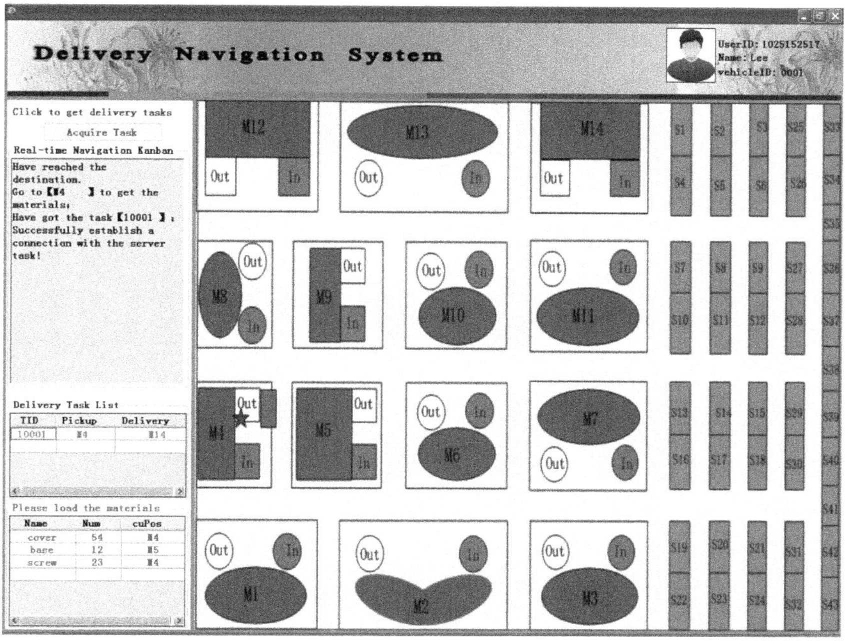

FIGURE 9.17 Monitor of parts loading.

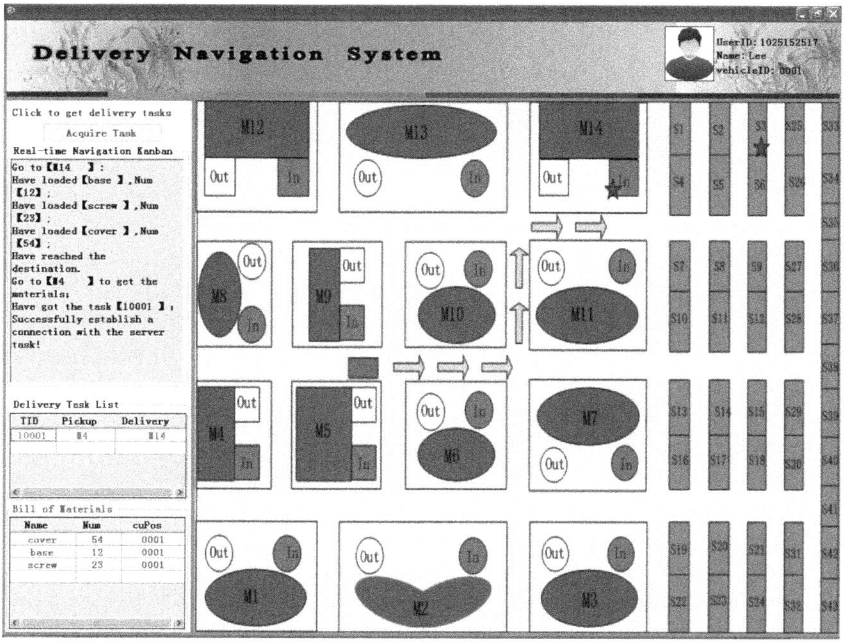

FIGURE 9.18 Navigating to the second picking-up site.

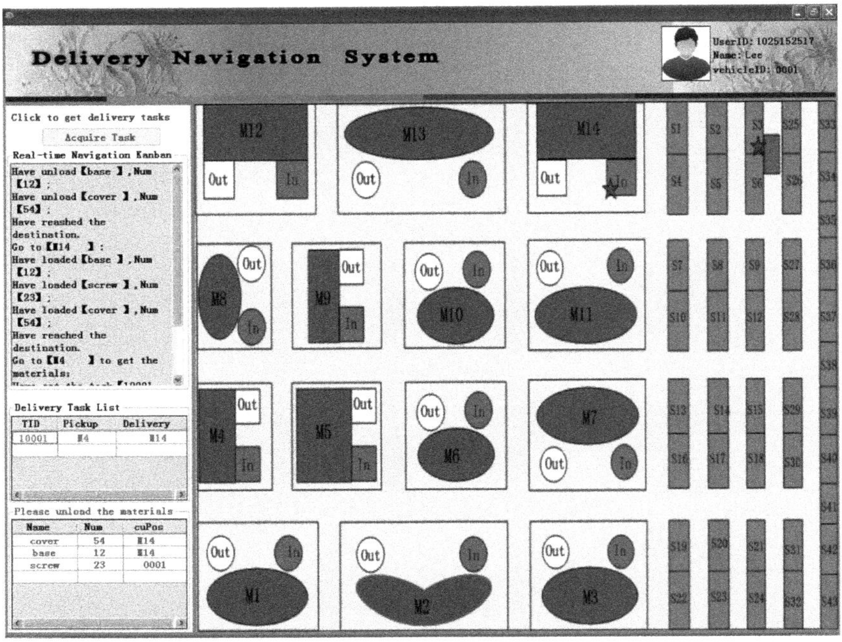

FIGURE 9.19 Real-time data-driven navigation of the intelligent trolley.

captured by the RFID readers on the vehicle. Thus, the row of the loaded parts will turn green, so that the manager and the operators are aware of the loading progress in real time.

After loading all the required materials, the system will provide instructions to the operator to reach the next location for picking up as shown in Fig. 9.18.

When the vehicle reaches the unloading location after picking up all the required materials, the system, similarly, will tell the operator which rack should the materials be put by showing another diagram of the storage rack. Once the item leaves the trolley, the leaving event will be sensed, and this item will resume red, indicating it is no longer on the vehicle. After finishing this task (Task TID 10001), the statuses of the vehicle becomes idle again, and can obtain another task from the system.

Fig. 9.20 shows the results of comparison for the traditional material handling method and the method proposed in Chapter 6. Fig. 9.20A shows the comparison of load/no-load ratio within the four vehicles from run distance. Fig. 9.20B shows the load/no-load comparison of each vehicle. From the comparison, we can see that the no-load ratio of all the trolleys is 19.6% lower as seen on the left of Fig. 9.20A. The total run distance of all trolleys for finishing all the tasks is 668 units shorter as seen on the right of Fig. 9.20A. The run distance of each trolley without loading is 195 (Vehicle 1), 56.5 (Vehicle 2), 175 (Vehicle 3), 183.5 (Vehicle 4) units shorter as seen in Fig. 9.20B.

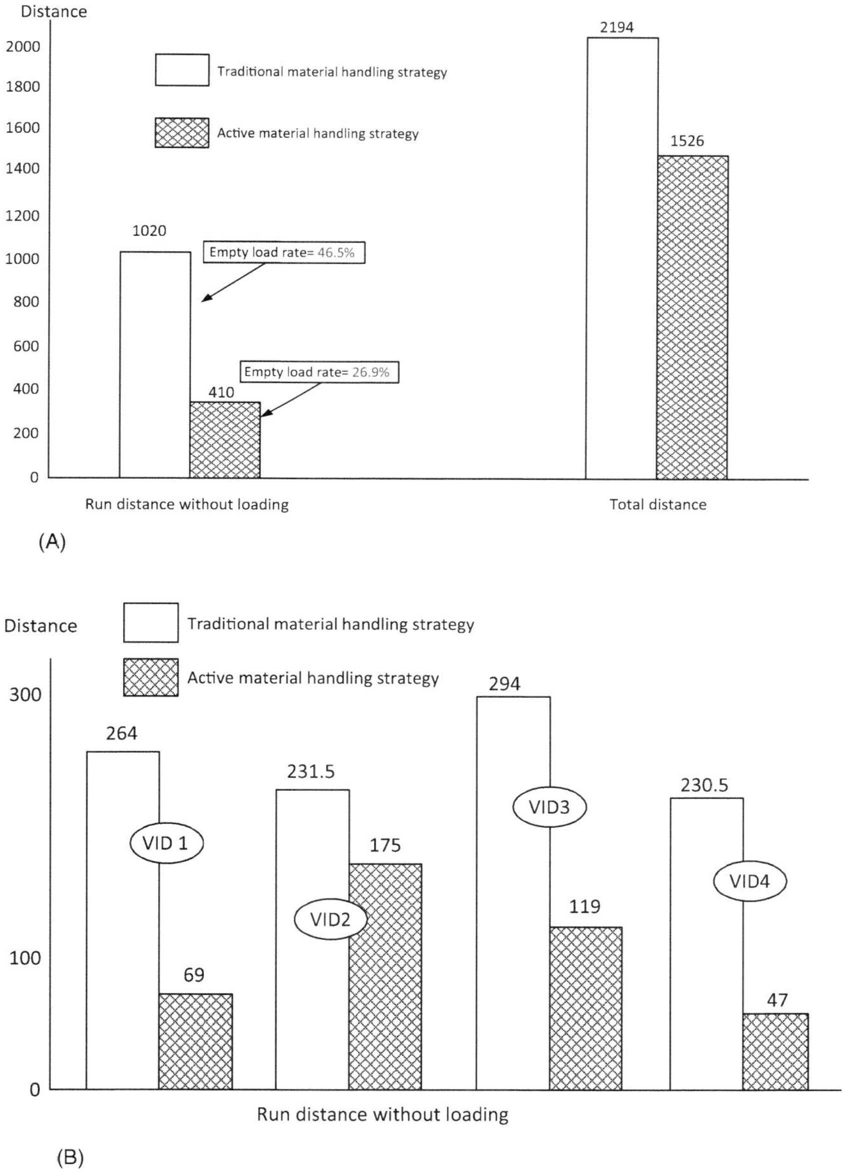

FIGURE 9.20 Results analysis and discussion. (A) Load/no-load comparison of all trolleys and (B) load/no-load comparison of each trolley.

9.7 IoT-ENABLED SMART STATION

9.7.1 The Case Scenario

This section shows that how the IoT-enabled station works in manufacturing systems. Four particular stations are studied here. As shown in Fig. 9.21, Station A and C are lathes. Station B is a gear-hobbing machine. Station D is an assembly station. Each station has an initial job queue from the scheduling system. Take Station B as an example. Currently, the first unprocessed job of the job list at Station B is J_1, and the succeeded process of J_1 will be processed at station C. The fourth unprocessed job of the job list at Station B is J_4, and the preceded process of J_4 is waiting to be processed at station D. In a similar way, the preceded and succeeded processes of other jobs included in the queue of station B will be processed at other stations, but they are not shown in Fig. 9.21.

9.7.2 Operation Guidance From the System

During the manufacturing execution stage, the real-time visibility explore at station B is used to reflect the real-time status and to show the related information, as shown in Fig. 9.22. On the left side of the interface, raw materials that are coming to this station are displayed. This is achieved by checking the radio signals from antennas installed at the material entrance. The RFID tags attached with products contain the information of the product ID. We can find that in this case, 94 end cups, 20 covers, 94 gears, etc. have been placed at the station. The task information is listed in the lower middle part. Currently, there are three tasks waiting to be processed. The Task JSQ011 is being processed at this particular time. The required batch for this task is 18, and 7 products have been finished. No disqualified products are found for this task, so the qualification rate is 100%. The due time of the current task is 18.03, and 38% of work has been finished till now. Real-time events are also displayed above the task information. The system will count the number of qualified products and scraps. Also, the status of the upstream and the downstream station is also available in the lower right corner. In the central part of the screen, operation guidance will be provided dynamically according to the Petri net model established in Chapter 4.

9.7.3 Real-Time Queuing Under Exceptions

The real-time queuing service for smart stations will requeue the job list based on the shared real-time information according to the designed algorithm [5] in Section 4.3.

For example, at time t, the current information of the jobs at station B and the current information of the relevant jobs of station A, C, and D are shown in Tables 9.14 and 9.15, respectively. The notations can be found in Section 4.6.

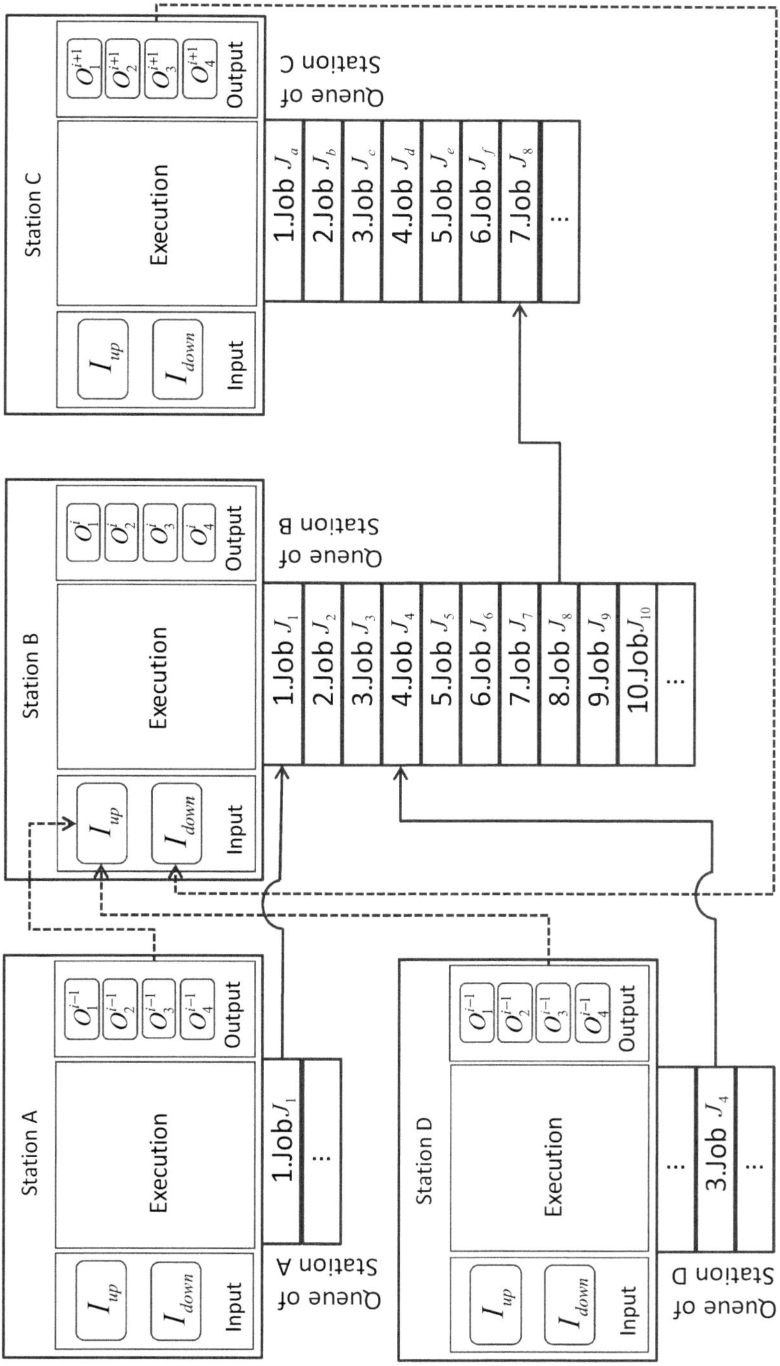

FIGURE 9.21 The relationships of stations A, B, C, and D.

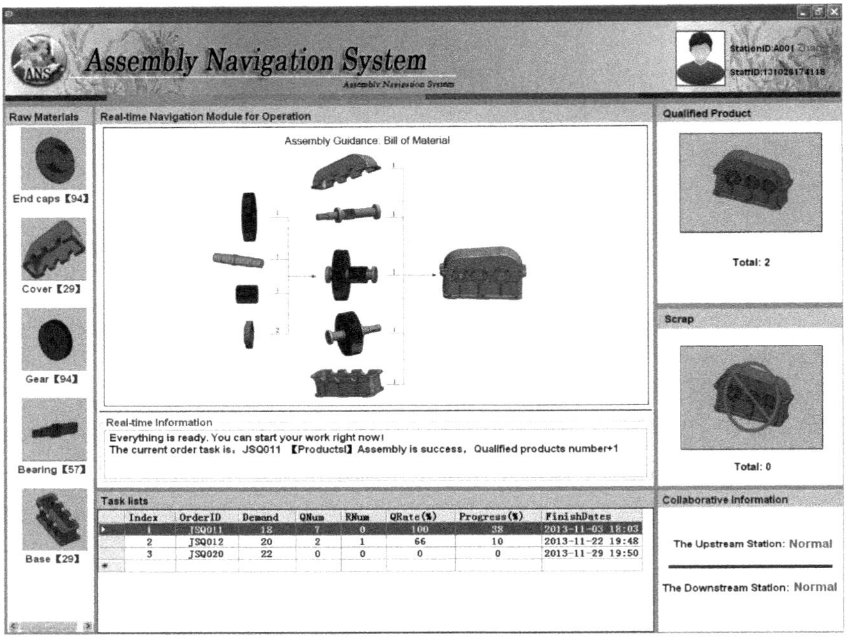

FIGURE 9.22 User interface of the operation guidance for smart stations.

TABLE 9.14 Information of the Jobs at Station B

Station	Process	p_j^i	ST_j^{i+1}	ET_j^{i-1}	$M_{i=1}$	M_{i+1}
Station B	J_1^i	30	25	10	Station A	
	J_2^i	50	70	25		
	J_3^i	20	100	90		
	J_4^i	25	130	80	Station D	
	J_5^i	35	145	115		
	J_6^i	35	200	150		
	J_7^i	40	220	160		
	J_8^i	50	270	230		Station C
	J_9^i	30	315	265		
	J_{10}^i	45	380	345		

TABLE 9.15 Current Information About the Jobs at Stations A, C, and D

Station	Process	p_j^{i-1}	ST_j	ET_j	Process status	Notes
Station A	J_1^{i-1}	25	2	0	Processing	Here ST_j represents the starting time of the following process. ET_j represents the completion time of the previous process.
	...					
Station D	...					
	J_4^{i-1}	30	110	50	Unprocessed	
Station C	...					
	J_8^{i+1}	30	315	295	Unprocessed	

TABLE 9.16 Exceptions Occurred Among the Related Upstream and Downstream Stations of Station B

Exceptions	Results
An unplanned machine fail occurred at the station A	20 min are required to fix the machine. The finished time of process J_1^{i-1} will be extended to 30 min
Material shortages of job J_4	The finished time of process J_4^{i-1} will be extended for 35 min
New insert job at station C	The start time of process J_8^{i+1} is delayed for 50 min

At the station B, those operations $J_x^i (x = 1 \sim 10)$ will be processed in the order of $\{J_1^i, J_2^i, J_3^i, J_4^i, J_5^i, J_6^i, J_7^i, J_8^i, J_9^i, J_{10}^i\}$. The tardiness penalty per unit time of each job is given as $w_{1\sim 10} = (2,1,3,2,3,4,3,1,2,2)$.

In this case, three exceptions listed in Table 9.16 are designed to occur among the related upstream and downstream stations of station B.

When the exceptions occur, the due times of the relevant unprocessed jobs at station B are changed, and the queuing service will update the new start time of the affected jobs as shown in Table 9.17.

Then, the queuing service will start to requeue the order of the unprocessed jobs of the job list at station B. The initial solution and the end condition are two important decisions in the application of Tabu Search (TS) algorithm since they directly affect the utilization of the algorithm. The initial solution is generated according to the size of the due date of each job. According to the rules of the earliest due date (EDD), the initial solution proposed is suitable for the algorithm. The maximum number of iteration is set as the end condition of this proposed algorithm for the sake of simplicity. According to (4.2), the initial solution is $q_k = \{J_1^i, J_2^i, J_3^i, J_4^i, J_5^i, J_6^i, J_7^i, J_9^i, J_8^i, J_{10}^i\}$. Through the proposed algorithm, the queue is optimized. The globally optimal

TABLE 9.17 Information of Jobs After Exceptions

Station	Process	p_j^i	ST_j^{i+1}	ET_j^{i-1}	$M_{i=1}$	M_{i+1}
Station B	J_1^i	30	25	30	Station A	
	...					
	J_4^i	25	130	115	Station D	
	...					
	J_8^i	50	320	230		Station C
	...					

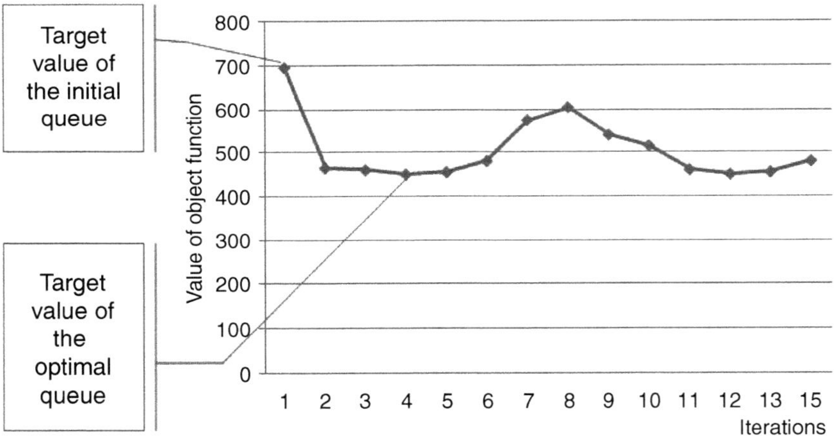

FIGURE 9.23 Iteration process of the TS algorithm.

solution is $q_o = \{J_2^i, J_1^i, J_3^i, J_5^i, J_6^i, J_4^i, J_7^i, J_9^i, J_8^i, J_{10}^i\}$. Fig. 9.23 shows the iteration process of the TS algorithm. The algorithm finds the optimal queue whose target value is "450" at the fourth iteration. But it still searches for other solutions until it reaches the end condition. It is obvious that the new queue is better, which is "295" shorter than the target value of the old queue ($\{J_1^i, J_2^i, J_3^i, J_4^i, J_5^i, J_6^i, J_7^i, J_8^i, J_9^i, J_{10}^i\}$).

9.8 REAL-TIME MANUFACTURING INFORMATION TRACK AND TRACE

The real-time information track and trace system is essential in the analysis of system performance and can provide necessary information for the requeuing service under exceptions. Fig. 9.24 illustrates the deviation

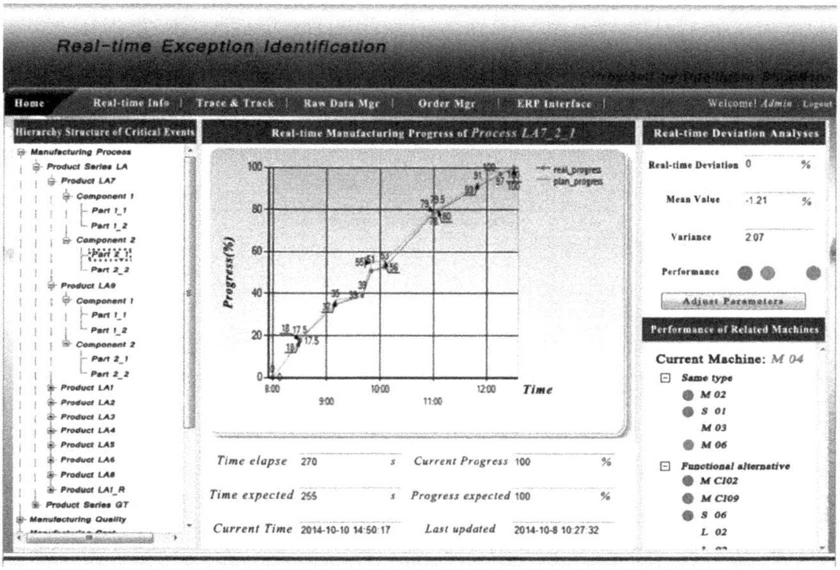

FIGURE 9.24 **Monitor of production deviation.**

between manufacturing plans and the actual processes. The projected progress and the real progress are represented by two curves in different colors. The hierarchy structure of the critical events are listed on the left. These events include those related to the progress, the production quality, the cost, etc. When clicking on different items in the list, the corresponding plans and actual situations will be displayed in the central chart. The summary of the selected process is also available under the chart, which includes the total production time of this product, current progress, the planned schedule, etc. The instant deviation, mean value, and variance of the deviation are calculated as well, which is shown on the upper right corner. Depending on the calculations, four different colors are used to mark the production process: green for the normal production (minor delays are allowed), yellow for the moderate delay, red for a major delay or other emergencies, and gray for the tasks that haven't started yet or has been finished for a long time.

The real-time manufacturing information can be viewed in different forms. In Fig. 9.25, The BOM of current product is shown. The two numbers below each component represent the planned number and the actual number of this part. For example, "GSZ101" is the ID for high-speed axles. The planned production number is 20, and only 10 have been produced now.

Similarly, in Fig. 9.26, the real-time information of products/parts or the equipment can be viewed through data sheets.

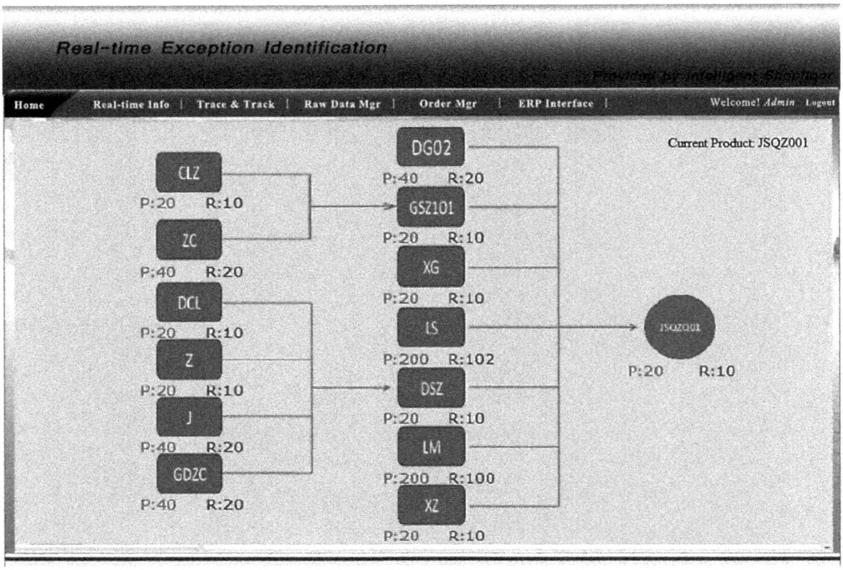

FIGURE 9.25 Monitor of BOM.

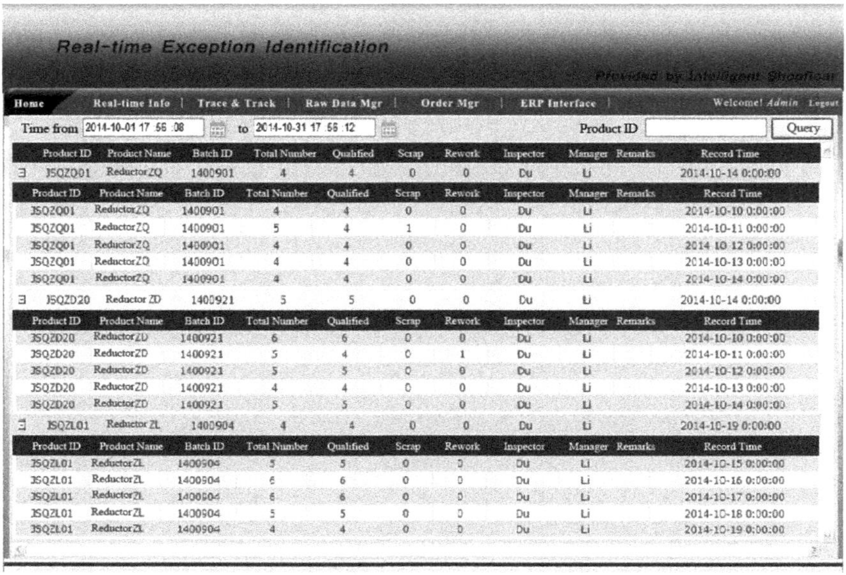

	Product ID	Product Name	Batch ID	Total Number	Qualified	Scrap	Rework	Inspector	Manager	Remarks	Record Time
⊟	JSQZQ01	ReductorZQ	1400901	4	4	0	0	Du	Li		2014-10-14 0:00:00
	Product ID	Product Name	Batch ID	Total Number	Qualified	Scrap	Rework	Inspector	Manager	Remarks	Record Time
	JSQZQ01	ReductorZQ	1400901	4	4	0	0	Du	Li		2014-10-10 0:00:00
	JSQZQ01	ReductorZQ	1400901	5	4	1	0	Du	Li		2014-10-11 0:00:00
	JSQZQ01	ReductorZQ	1400901	4	4	0	0	Du	Li		2014 10 12 0:00:00
	JSQZQ01	ReductorZQ	1400901	4	4	0	0	Du	Li		2014-10-13 0:00:00
	JSQZQ01	ReductorZQ	1400901	4	4	0	0	Du	Li		2014-10-14 0:00:00
⊟	JSQZD20	Reductor ZD	1400921	5	5	0	0	Du	Li		2014-10-14 0:00:00
	Product ID	Product Name	Batch ID	Total Number	Qualified	Scrap	Rework	Inspector	Manager	Remarks	Record Time
	JSQZD20	ReductorZD	1400921	6	6	0	0	Du	Li		2014-10-10 0:00:00
	JSQZD20	ReductorZD	1400921	5	4	0	1	Du	Li		2014-10-11 0:00:00
	JSQZD20	ReductorZD	1400921	5	5	0	0	Du	Li		2014-10-12 0:00:00
	JSQZD20	ReductorZD	1400921	4	4	0	0	Du	Li		2014-10-13 0:00:00
	JSQZD20	ReductorZD	1400921	5	5	0	0	Du	Li		2014-10-14 0:00:00
⊟	JSQZL01	Reductor ZL	1400904	4	4	0	0	Du	Li		2014-10-19 0:00:00
	Product ID	Product Name	Batch ID	Total Number	Qualified	Scrap	Rework	Inspector	Manager	Remarks	Record Time
	JSQZL01	ReductorZL	1400904	5	5	0	0	Du	Li		2014-10-15 0:00:00
	JSQZL01	ReductorZL	1400904	6	6	0	0	Du	Li		2014-10-16 0:00:00
	JSQZL01	ReductorZL	1400904	6	6	0	0	Du	Li		2014-10-17 0:00:00
	JSQZL01	ReductorZL	1400904	5	5	0	0	Du	Li		2014-10-18 0:00:00
	JSQZL01	ReductorZL	1400904	4	4	0	0	Du	Li		2014-10-19 0:00:00

FIGURE 9.26 Monitor of real-time product information.

9.9 REAL-TIME KEY PRODUCTION PERFORMANCES MONITOR MODULE

9.9.1 Details of the Case

In our case, the production process can be divided into three stages: the part processing stage, the component assembly stage, and the final assembly stage. While the manufacturing of many parts and accessories is outsourced to suppliers and shareholding subsidiaries, there are two main assembly lines (Assembly 1 and Assembly 2) and manufacturing processes for three main parts in the shop floor. Two of them are for Part 11 (Part 111 and Part 112) and one for Part 12. Among the different manufacturing areas, there are three kinds of material handling procedures: raw material distribution (from raw material area to part processing line buffer), WIP circulating process (from part processing line to part assembly line, and part assembly line to final assembly line), and finished product storing process (product assembly line to finished product area). There is a quality inspection site in the shop floor as well. Take a processing and assembly event as an example; each process checks its operator and material status before the process. If both of them are ready, a process operates according to a schedule.

9.9.2 The Hierarchy Timed Color Petri Net Model

According to the manufacturing routes, a hierarchy timed color Petri net (HTCPN) model is built up in CPN tools [6] as seen in Fig. 9.27. There is one overall model for the critical event and it has six macrotransitions, which are linked to six complex event sub-CPN models. The six subnet models represent six production processes distributed in six different locations—Line Assembly 1, Line Part 11, Line Part 12, Line Part 111, Line Part 112, and Line Part 121, respectively. Similarly, each complex event model has several macrotransitions, which are linked to the basic event sub-CPN models. To simplify the presentation, only the subnet for Line Assembly 1 and the subnet for the process and assembly event in Line Assembly 1 are given. The global color set declarations are given in Table 9.18. The firing time and functions are given in the model shown in Fig. 9.27 referring to the collected data from the case company.

From the overview of the HTCPN model in Fig. 9.27A, there is a piece of material in in-buffer Part 111 (Place MPart111); value 0 after symbol "@" indicates its time stamp. The token in this subnet is of DATAt color type, which indicates that the job ID is a string type. When the model starts running, tokens move from one place to another, and in each place one gets a certain time stamp for each token. Tokens enter the first subnet Part 111 when the line uses the material.

Details of the six macrotransitions for each processing line are presented. For example, the macrotransition for Assembly 1 is given in Fig. 9.27B. Similarly,

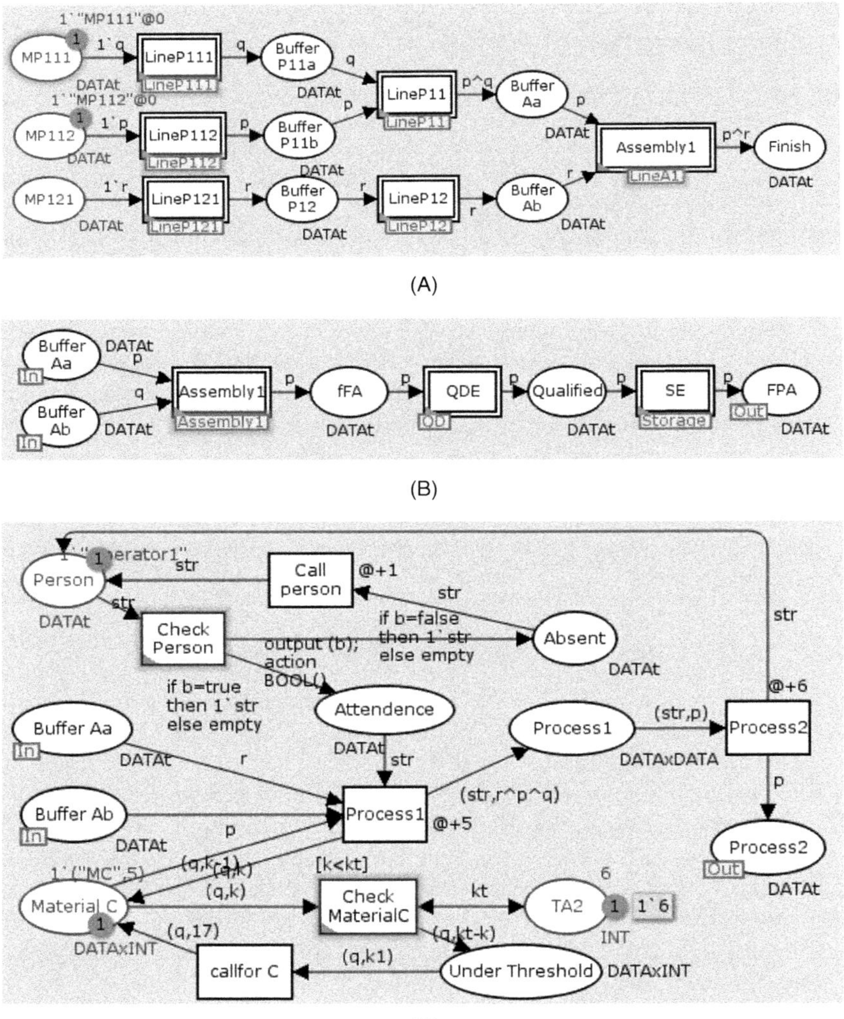

FIGURE 9.27 Hierarchy timed color Petri net (HTCPN) model for case shop floor. (A) Overview HTCPN model for the case shop floor, (B) sub-HTCPN in Line Assembly 1, and (C) sub-HTCPN in assembly process.

the detail events of the three macrotransitions for Line Assembly 1 are given as assembly process event (Transition Assembly), quality detection event (Transition QDE), and storage event (Transition SE).

As seen in Fig. 9.27C, the assembly process event is used to exemplify the model. When the assembly process should be executed according to the planned

TABLE 9.18 Global Color Set Declarations

Colset INT=int; Colset INTt=int timed;
Colset DATA=string; Colset BOOL=bool;
Colset INTxDATA=product INT*DATA;
Colset State=with well|bad|scrap timed;
Colset DATAt=DATA timed;
Colset StatexDATA=product State*DATAt;
Colset StatexDATAxDATA=product State*DATA*DATA;
Colset DATAxINT=product DATA*INT timed;
Colset DATAxBOOL=product DATA*BOOL;
Colset DATAxDATA=product DATA*DATA timed;
Colset DATAxDATAxBOOL=product DATA*DATA*BOOL;
Colset DATAxDATAxDATA=product DATA*DATA*DATA timed;
Var n,i,j,k,it,jt,kt,i1,j1,k1:INT; Var wait:INTt;
Var p,q,r:DATA;Var s:State;Var str:DATAt;Var b:BOOL

time, the PN model starts running simultaneously. At the beginning of each assembly process, the model checks the status of operators and the materials in advance. It is stated that the real-time manufacturing data is captured by RFID antennas installed at different areas. The status of tokens is updated accordingly. The tokens for an operator (Place Person) and WIPs (Place Assembly1a and Assembly1b) are also DATAt type. The token for material C is represented as (DATA, INT), where DATA is material C's ID and INT is its quantity. After checking the status, the token for person goes into Place Attendance or Place Absent, which indicates that the person is ready or unready. If the WIPs are unready, the process needs to wait. Besides, the material C's condition is always checked. If the number is under the threshold *k*, the material scheduler (Place Under Threshold) receives a token with attribute (DATA, INT) that presents the material ID and the shortage status. Then, the material is replenished accordingly. In this case, once all the operators, WIPs, and materials are ready, the process starts to operate according to the plan for assembly. Moreover, the delay time (Exception) between the trigger time and planned time can be easily calculated, and this exception event can be sent to the upper-level management system in time.

Once the HTCPN model is constructed, the production performance acquirement measures are performed to analyze the real-time production performance. Fig. 9.28 shows the results of performance analysis for several factors for the case, including quality distribution, cycle time, real-time progress, and the cost of the manufacturing process.

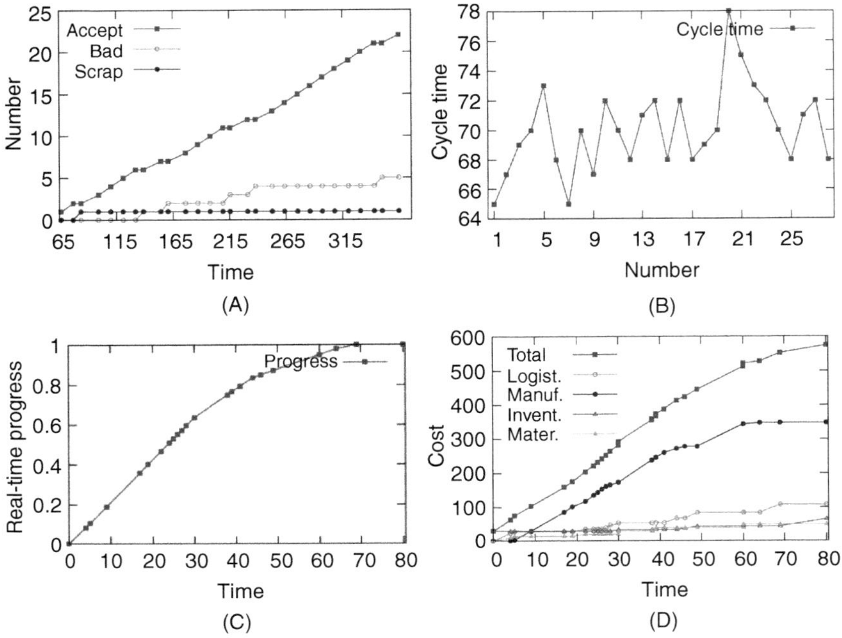

FIGURE 9.28 Simulation results for the case shop floor. (A) Quality distribution, (B) cycle time, (C) production progress, and (D) production cost.

REFERENCES

[1] Siemens, "Tecnomatix." Available from: https://www.plm.automation.siemens.com/en_us/products/tecnomatix/

[2] Y. Zhang, D. Xi, R. Li, S. Sun, Task-driven manufacturing cloud service proactive discovery and optimal configuration method, Int. J. Adv. Manuf. Technol. 84 (1–4) (2015) 29–45.

[3] Y. Zhang, G.Q. Huang, S. Sun, T. Yang, Multi-agent based real-time production scheduling method for radio frequency identification enabled ubiquitous shopfloor environment, Comput. Ind. Eng. 76 (1) (2014) 89–97.

[4] Y. Zhang, G. Zhang, W. Du, J. Wang, E. Ali, S. Sun, An optimization method for shopfloor material handling based on real-time and multi-source manufacturing data, Int. J. Prod. Econ. 165 (2015) 282–292.

[5] Y. Zhang, J. Xu, S. Sun, T. Yang, Real-time information driven intelligent navigation method of assembly station in unpaced lines, Comput. Ind. Eng. 84 (2015) 91–100.

[6] Y. Zhang, W. Wang, N. Wu, C. Qian, IoT-enabled real-time production performance analysis and exception diagnosis model, IEEE Trans. Autom. Sci. Eng. 13 (3) (2015) 1–15.

Chapter 10

Conclusions and Future Works

10.1 CONCLUSIONS

Internet of things (IoT) provides the potentiality to collect and communicate real-time data within the manufacturing systems, thereby achieving dynamic optimization and control of production. This helps to develop a more intelligent manufacturing system with higher flexibility and transparency. By extending IoT technologies to manufacturing field, this book proposes and develops an overall architecture and solution for IoT-MS. Under this architecture, the core models and technologies such as smart model of bottom manufacturing resources, and so on are discussed in details. The main contributions of this book could be summarized as follows.

1. Propose an overall architecture and solution for optimization of IoT-MS
   Considering the advantages of the IoT, an overall architecture of the manufacturing systems using IoT (IoT-MS) is presented. It aims to provide a new paradigm for real-time monitor, analysis, control, and dynamic optimization of the manufacturing systems by extending the IoT technologies to manufacturing field. Under this IoT-MS architecture, the manufacturing things such as operators, machines, pallets, materials, and so on can be embedded with sensors; they can interact with each other. The changed information and their status could thus be tracked and integrated with heterogeneous enterprise management information systems. The proposed IoT-MS will facilitate the real-time information driven active monitor, analysis, control, and dynamic optimization of the manufacturing systems.
2. Design a real-time and multisource manufacturing information sensing system
   In order to capture the real-time status of the distributed manufacturing resources, a real-time and multiple-source manufacturing information sensing system (RMMISS) is designed. The key technologies, such as deployment of multiple sensors, sensor manager, and multisource manufacturing information processing and sharing are discussed in detail. By implementing the proposed RMMISS, smart sensors can be managed in a "plug-and-play" fashion, and the real-time and multiple-source manufacturing information can be correctly captured and actively visited.

Optimization of Manufacturing Systems Using the Internet of Things
http://dx.doi.org/10.1016/B978-0-12-809910-0.00010-X

3. Design a smart model for IoT-enabled assembly station
   To enhance the intelligence of the bottom manufacturing resources, the overall architecture of a smart assembly station is designed by adopting RFID technology. Under this architecture, four core molds, namely real-time status monitoring, real-time production guiding, real-time production data sharing and real-time production requeuing, are designed to provide the optimization of navigation for the assembly stations. Then, the real-time information of manufacturing resources among assembly stations and assembly exception could be sensed and captured. The real-time manufacturing information sharing could also be achieved. The real-time production requeuing algorithm based on Tabu Search is designed to optimize assembly processing of smart assembly stations. The designed architecture will contribute to the real-time information driven process monitor and control between the manufacturing system and stations.
4. Design a cloud computing based manufacturing resources configuration method
   Based on the smart station, an overall architecture of cloud computing based manufacturing resources configuration is proposed to realize the full sharing, high collaboration, and flexible configuration of manufacturing resources. Under this architecture, a cloud machine model is built considering both the static and dynamic manufacturing information, and then published into the CMfg platform. The task-driven service proactive discovery mechanism promotes the providers' initiative. Consequently, the efficiency of service discovery is highly enhanced. Through the designed grey relational analysis (GRA)-based evaluation method, the service optimal selection and composition are performed.
5. Design a smart model for IoT-enabled assembly trolley
   A novel material handling model is proposed to improve the decision making based on the real-time and multisource manufacturing information. Comparing with existing central material handling method, the proposed method allows each trolley to request the move task actively, and each trolley could get the optimal move tasks according to their real-time status. Three key technologies are designed and developed to implement the presented material handling model. They are IoT-enabled smart trolley, real-time information exchange mechanism, and combination optimization method for material handling tasks.
6. Design a real-time key production performances analysis method
   As the amounts of distributed manufacturing data obtained by the sensors are often meaningless and cannot be used by the upper level managers directly, an overall architecture of real-time key production performance analysis method (PPAM) are designed to address this problem. First, the IoT devices are extended to manufacturing field to dynamically sense manufacturing data. Then, the multilevel event model and hierarchy timed color Petri net are used to integrate the discrete raw manufacturing data into critical events.

At last, the Decision Tree technologies are applied to extract the production anomalies and find the causes of anomalies.

7. Design a real-time information driven production scheduling strategy and method
A real-time information driven production scheduling strategy and method is designed to bridge the gap between production planning and control. The real-time information-driven production system consists of four modules. They are respectively as equipment agent, capability evaluation agent, real-time scheduling agent, and production execution monitor agent. Equipment agent is employed to capture and process real-time information in the shop floor. In the process planning stage, capability evaluation agent is used to accomplish the optimal tasks allocation according to the real-time utilization ration of each machine. As the name suggests, real-time scheduling agent is responsible for the manufacturing tasks scheduling or rescheduling according to the traced real-time information. Production execution monitor agent aims to track and trace the real-time status of different manufacturing processes.

10.2 FUTURE WORKS

Future works of our research group will involve the following aspects.

1. New mode and decision strategy for next advanced manufacturing systems.
2. Cyber-physical system based smart manufacturing resource modeling.
3. Big data based product life cycle management for sustainable manufacturing.
4. Real-time optimization strategy and algorithms for manufacturing systems.
5. Collaborative optimization model and method for intelligent production-logistics system.

Index

C

D

E

T

U

V

W

X

CPI Antony Rowe
Chippenham, UK
2017-03-03 22:44